REPORT

Energy Services Analysis

An Alternative Approach for Identifying Opportunities to Reduce Emissions of Greenhouse Gases

Keith Crane • Liisa Ecola • Scott Hassell • Shanthi Nataraj

Sponsored by the Office of Energy Efficiency and Renewable Energy of the U.S. Department of Energy

Environment, Energy, and Economic Development

A RAND INFRASTRUCTURE, SAFETY, AND ENVIRONMENT PROGRAM

This report was sponsored by the United States Department of Energy and was conducted in the Environment, Energy, and Economic Development Program within RAND Infrastructure, Safety, and Environment, a division of the RAND Corporation.

Library of Congress Cataloging-in-Publication Data

Energy services analysis : an alternative approach for identifying opportunities to reduce emissions of greenhouse gases / Keith Crane ... [et al.].
  p. cm.
 Includes bibliographical references.
ISBN 978-0-8330-6036-5 (pbk. : alk. paper)
1. Energy auditing. 2. Energy conservation. 3. Greenhouse gas mitigation. I. Crane, Keith, 1953-

  TJ163.245.E54 2012
  333.791'6—dc23

         2012010449

The RAND Corporation is a nonprofit institution that helps improve policy and decisionmaking through research and analysis. RAND's publications do not necessarily reflect the opinions of its research clients and sponsors.

RAND® is a registered trademark.

Published 2012 by the RAND Corporation
1776 Main Street, P.O. Box 2138, Santa Monica, CA 90407-2138
1200 South Hayes Street, Arlington, VA 22202-5050
4570 Fifth Avenue, Suite 600, Pittsburgh, PA 15213-2665
RAND URL: http://www.rand.org/
To order RAND documents or to obtain additional information, contact
Distribution Services: Telephone: (310) 451-7002;
Fax: (310) 451-6915; Email: order@rand.org

Preface

Efforts to conserve energy and reduce emissions of associated greenhouse gases often focus on improving the efficiency of current technologies. However, in many instances, changing the ways in which a service is provided may open up greater opportunities to reduce energy use than attempting to make current ways of delivering the service more efficient. This report elaborates on and demonstrates energy services analysis, an analytical approach that investigates opportunities for reducing energy use by changing the manner in which a service is provided, rather than assessing means of making existing technologies more energy efficient, to reduce energy consumption. It looks in-depth at two case studies: substituting electronically disseminated newspapers for paper copies and substituting vehicle-sharing services for personally owned vehicles.[1]

This research was requested by the Office of Energy Efficiency and Renewable Energy of the U.S. Department of Energy.

This report should be of interest to researchers in the field of energy conservation looking for alternative means of analysis. It should also be of interest to researchers and companies engaged in (1) printing and publishing and (2) vehicle sharing—the two service categories that we analyze. It should also be of interest to federal agencies that wish to promote policies to conserve energy.

Other RAND reports that address energy conservation include *Impacts on U.S. Energy Expenditures and Greenhouse-Gas Emissions of Increasing Renewable-Energy Use* (Toman, Griffin, and Lempert, 2008), *Improving the Energy Performance of Buildings* (Ries, Jenkins, and Wise, 2009), and *Integrating U.S. Climate, Energy, and Transportation Policies: Proceedings of Three Workshops* (Ecola et al., 2009).

The RAND Environment, Energy, and Economic Development Program

This research was conducted in the Environment, Energy, and Economic Development Program (EEED) within RAND Infrastructure, Safety, and Environment (ISE). The mission of ISE is to improve the development, operation, use, and protection of society's essential physical assets and natural resources and to enhance the related social assets of safety and security of individuals in transit and in their workplaces and communities. The EEED research portfolio addresses environmental quality and regulation, energy resources and systems, water resources

---

[1] This report uses the term *vehicle sharing* to emphasize the fact that one benefit is the ability to share both cars and trucks. The more common term is *car sharing*.

and systems, climate, natural hazards and disasters, and economic development—both domestically and internationally. EEED research is conducted for government, foundations, and the private sector.

Questions or comments about this report should be sent to the project leader, Keith Crane (Keith_Crane@rand.org). Information about EEED is available online (http://www.rand.org/ise/environ). Inquiries about EEED projects should be sent to the following address:

Keith Crane, Director
Environment, Energy, and Economic Development Program, ISE
RAND Corporation
1200 South Hayes Street
Arlington, VA 22202-5050
703-413-1100, x5520
Keith_Crane@rand.org

Contents

Tables

Summary

The purpose of this technical report is to identify and evaluate new means to reduce energy use and greenhouse gas (GHG) emissions by employing energy services analysis (ESA). Most efforts in this area focus on ways to reduce energy use and GHG emissions by making existing processes more efficient. This report uses ESA to examine possibilities for changing how a service is delivered to reduce energy use and GHG emissions. The report introduces ESA, explains how it differs from conventional approaches and how this type of analysis can be conducted, uses an ESA framework to analyze how changes in the provision of two common services might result in reductions in energy use and GHG emissions, suggests other areas in which ESA could be applied, and ends with some thoughts on using ESA more broadly.

Energy Services Analysis

Energy consumption and GHG emissions in the United States are generally analyzed by end use: residential, commercial, industrial, and transportation. These categories focus on the user, not necessarily the purpose for which the energy is consumed. Stemming from this focus on end use, efforts to identify ways to conserve energy often target improving the efficiency of current technologies. However, in many instances, changing the way in which a service is provided can open up more or better opportunities to reduce energy use than attempting to make the current way of delivering the service more efficient. ESA evaluates opportunities for reducing energy use and GHG emissions by changing how a service is provided.

ESA seeks to rectify two of the main drawbacks of focusing on energy end use rather than the provision of the service desired. First, focusing on the end use does not address the purpose for which the energy is consumed. Second, focusing on the direct use of the energy ignores indirect consumption. For example, the owner of a car purchases gasoline, but energy is also consumed in producing the materials, such as steel, used to manufacture the car, in assembling the car, delivering the vehicle, and the other stages of production.

ESA addresses both of these issues. First, ESA focuses on the service that the energy is used to provide rather than on the consumer of the energy. The goal of ESA is to determine whether the service could be provided in another way that requires less energy. Second, ESA reallocates energy use from the purchaser or consumer to the individual service that is consumed. In so doing, ESA includes all energy used in providing a service, whether directly (by the end user) or indirectly (at any other point in the chain of service provision). In the example above, ESA allocates the energy required to manufacture, distribute, and drive the car to the service of mobility, provided by the vehicle, rather than distribute the energy used by the driver, the automobile manufacturer, and the steel mill into different categories.

Applying Energy Services Analysis

ESA can be conducted using either of two analytical methods. In process analysis, the entire process of creating a product or service, from raw materials to disposal, is laid out. The energy used at each stage of the process is estimated and summed to provide an estimate of total energy consumption, direct and indirect. Process analysis requires highly disaggregated data and is generally conducted at the firm level because it looks only at the manufacturing process. Life-cycle assessment differs from process analysis in that it evaluates energy use and GHG emissions from cradle to grave. Life-cycle assessment measures all the energy consumed, from the equipment and raw materials used to manufacture the product, to production, to final use and disposal.

ESA involves creating a framework for thinking about alternative means of providing a service. In this report, we use a typology of human wants and needs developed by Costanza et al. (2007) to classify various services so as to evaluate energy consumption by competing means of providing these services. These wants and needs are subsistence, security, affection, understanding, participation, leisure, spirituality, creativity, identity, and freedom. All the energy that people use ultimately fulfills one or more of these wants or needs.

Because energy can be measured with different metrics, our application of ESA focuses on GHG emissions. This makes it easier to compare the effects across various sectors: delivery of written news, vehicle sharing, food, clothing, health care, and waste disposal.

Delivery of Written News

We conducted an ESA for the delivery of written news, a service that fulfills the need for access to information. Electronic dissemination of written news already substitutes for the delivery of some print newspapers. With the advent of electronic readers ("e-readers") and tablet computers, the shift from print to electronic dissemination appears set to accelerate.

Paper manufacturing, printing, and newspaper distribution release substantial amounts of GHGs. The frequency and volume of newspapers makes them particularly energy intensive. We estimate that, in the United States, one newspaper subscription releases 94.7 kg of carbon dioxide annually, for production, printing, and delivery (see Table 3.1 in Chapter Three). Providing written news through alternative means could potentially save substantial amounts of energy and reduce emissions of GHGs. In contrast, the production and operation of a single e-reader or tablet computer generates far fewer GHG emissions, assuming that emissions produced during the manufacture of these devices are spread out over a three-year product life span. Table S.1 shows the differences in GHG emissions from a single newspaper subscription and from using these devices.

We calculated potential reductions by disseminating written news with e-readers rather than newspapers in a "what-if" scenario (that is, what if each current newspaper subscription were replaced today with an e-reader or tablet computer). Adopting e-readers could reduce GHG emissions from publishing and distributing newspapers by 74 percent; using tablet com-

Table S.1
Annual Greenhouse Gas Emissions Associated with Reading a Newspaper and Reading the Same News on a Tablet Computer or E-Reader

Emission Type	Unit	Newspaper	Tablet Computer	E-Reader
GHG emissions per subscription	Kg of CO_2e	94.7	35	24.7
Newspaper subscriptions	Millions	45.9		
GHG emissions	Million metric tons	4.35	1.61	1.13
Reduction compared to newspapers	Million metric tons		2.74	3.21
	Percentage		63	74

NOTE: CO_2e = carbon dioxide (CO_2) equivalent.

puters could result in a 63 percent reduction, assuming that all the GHG emissions associated with producing and operating e-readers or tablet computers are ascribed to reading newspapers. If a more realistic assumption is adopted, that the emissions associated with these devices should be spread across other activities pursued on these devices, the difference would be on the order of 84 to 89 percent less, respectively.

Policy Options

Because of these energy savings and reductions in GHG emissions, the U.S. Department of Energy (DOE) may wish to ensure that consumers are made better aware of the trade-offs in energy use between these two different methods of obtaining written news. Options for increasing this awareness include the following:

- Encourage manufacturers to provide information on the energy efficiency of electronic readers and information and communication technology (ICT) equipment.
- Monitor technological developments in ICT network equipment and usage.
- Provide consumers with the right to retain access to information they have purchased so that consumers do not have to repurchase content if they switch or upgrade devices.

Mobility: Sharing, Rather Than Owning, Vehicles

We also used ESA to evaluate the provision of the service of mobility by shared vehicles rather than by personally owned vehicles. Mobility fulfills a wide variety of human wants and needs, ranging from subsistence to leisure. Vehicle sharing provides personal mobility without owning a vehicle. Participants in vehicle-sharing programs generally give up their personally owned motor vehicles or forgo vehicle purchases, using vehicles owned by the program when needed.

Most transportation in the United States takes place in personally owned light-duty vehicles. The number of miles that the owners of these vehicles drive each year increased steadily until 1999; since then, they have leveled off. Many observers think that one reason for these increases is that vehicle ownership has substantial fixed costs that are incurred regardless of how much the vehicle is driven; the extra costs of taking a trip in the vehicle are small. Vehicle sharing converts these fixed costs into variable costs by charging individuals per trip. Faced with higher out-of-pocket costs per trip, participants take fewer trips by motor vehicle, reducing overall energy use and GHG emissions. Moreover, drivers who do not drive much can save money through vehicle sharing because they do not have to pay the high fixed costs of purchasing and owning a vehicle.

Vehicle sharing is available in many U.S. cities. Current membership figures are about 560,000, which represents 0.27 percent of U.S. drivers. Vehicle-sharing services have been most successful in neighborhoods with good access to public transit and higher-than-average density. Former vehicle owners living in these neighborhoods can easily shift some of their trips to public transit or nonmotorized modes. Vehicle sharing provides efficient alternatives for other niche users, such as businesses and governments that operate vehicle fleets and students on college campuses.

Vehicle sharing has the potential to substantially reduce GHG emissions in three ways. First, some studies (see, e.g., Millard-Ball et al., 2005; and Martin and Shaheen, 2010) find substantial average reductions in vehicle miles traveled after a participant joins a vehicle-sharing organization. Second, shared vehicles are generally more fuel efficient than the existing vehicle fleet. Finally, because members drive fewer miles and vehicles are used more intensively, vehicle sharing reduces the number of cars needed, which, in turn, reduces energy used and GHGs emitted in the manufacture of vehicles. Using this analysis, we calculate that a single driver who shifts from personally owning a motor vehicle to participating in a vehicle-sharing program would likely emit 893 kg of CO_2e per year less than if he or she had continued to own and operate a vehicle. Although the average vehicle-sharing participant would reduce his or her GHG emissions by about 37 percent, it is important to note that the average vehicle-sharing participant drives far less than the average American driver.

We estimated potential reductions in GHG emissions under three cases, which reflect current vehicle-sharing programs. In the base case, we use the current penetration rate of 0.27 percent to calculate reductions in GHG emissions from participation in vehicle-sharing programs in that year. In the second case, called *supportive policies* because it assumes that policies that actively support vehicle sharing lead to increased participation, we estimate potential reductions if 4.5 percent of U.S. drivers were to participate in vehicle-sharing programs. Our third case estimates GHG reductions under the assumption that 12.5 percent of U.S. drivers aged 21 and up in major metropolitan areas join such programs, an upper-bound estimate of potential penetration rates from a study by Shaheen, Cohen, and Roberts (2006).

We found that participation in vehicle-sharing programs may currently result in reductions in GHG emissions of 0.5 million metric tons of CO_2e per year. By way of comparison, we find that light-duty vehicles accounted for 1,071 million metric tons of emissions in the United States in 2009, based on U.S. Energy Information Administration estimates of total GHG emissions from light-duty vehicles. In the second and third cases, the absolute amount of reductions in emissions of GHGs would be substantial, all other factors being equal (Table S.2). These assumptions are based on current models of vehicle sharing; however, other models that are not yet in widespread use (such as peer-to-peer vehicle sharing) may make vehicle sharing more attractive.

Policy Options

Encouraging the expansion of vehicle sharing or other alternatives to private ownership of motor vehicles could substantially reduce energy consumption and GHG emissions. However, these alternatives would have to be competitive in terms of convenience as well as cost. Vari-

Table S.2
Potential Reductions in Greenhouse Gas Emissions from Switching from Personally Owned Vehicles to Vehicle Sharing, Under Three Cases

Factor	Base Case	Supportive Policies Case	Maximum Adoption Case
Number of participants	560,000	7,500,000	20,300,000
Assumed percentage adoption by drivers in urban areas	0.27	4.5	12.5
Total reductions in GHG emissions (millions of metric tons)	0.5	6.7	18.1
Percentage reduction from GHG emissions from U.S. light-duty fleet	0.05	0.6	1.7

NOTE: Reductions in GHG emissions per participant are 893 kg. The percentage reduction is based on 1,071 million metric tons of GHG emissions from light-duty vehicles (EIA, 2012, Table 19).

ous policies might help speed the adoption of vehicle sharing as it currently exists, as well as promote new models:

- Adopt parking policies that allow one-way, dynamic vehicle sharing.
- Change insurance regulations to facilitate peer-to-peer vehicle sharing.
- Alleviate the tax burden on shared vehicles.
- Help advance technologies that facilitate vehicle sharing.
- Legalize and provide better operating environments for shared neighborhood vehicles.
- Research material advances to enable greater user of shared neighborhood vehicles.
- Develop better technology for ride-matching services to enable smart paratransit and dynamic ride-sharing.
- Continue research into and regulation of the technology to enable driverless vehicles.
- Change liability laws to enable use of driverless vehicles.

Other Opportunities for Employing Energy Services Analysis

Many other services could be evaluated for their potential to be provided in alternative ways. Some of these alternatives are available now, while others are not yet commercially available. Examples of other services that could be assessed with ESA include the following:

- Food: ESA can be used to evaluate the energy used to produce locally grown food as opposed to food produced elsewhere and shipped. Employing ESA to compare the energy and GHGs emitted in producing certain types of food, such as protein sources, could identify foods that entail the lowest amounts of energy. ESA could also be used to evaluate the energy intensity of food preparation, comparing lower-energy cooking techniques or out-of-home food preparation with traditional food preparation.
- Clothing: ESA could be used to evaluate energy savings associated with materials that last longer or need less laundering, or better channels for distributing used clothing to extend its useful life.

- Health care: ESA could be used to evaluate the life-cycle energy consumption associated with particular medical treatments. It could also look at ways to reduce energy consumption in health care settings or reduce the length of stay in such facilities so that more patients can be served.
- Waste disposal: ESA could be used to evaluate the energy consumption of landfills as opposed to recycling or other forms of disposal.

In the residential energy-use category, direct energy use accounts for only one-third of all energy consumed, while indirect energy accounts for two-thirds. ESA thus offers a practical way to look for energy savings by identifying the services that consume the most energy and investigating alternative means of delivering those services.

Conclusions

ESA provides a useful means of looking at energy use differently. Changing the way in which a service is delivered can translate into large reductions in energy use and GHG emissions. Reductions ought to be considered both per user and in aggregate, as alternative ways of providing a service result in aggregate declines stemming from widespread substitution of the new activity for the old. Table S.3 summarizes these differences in per capita and aggregate decreases for both of our examples, as well as our assumptions about the diffusion of the alternative way of providing the service.

For the delivery of written news, the percentage reduction estimated in this summary is as much as 89 percent, both on a per-user basis and in aggregate, assuming that all users switch from newsprint to reading electronically. The absolute reductions could be as high as 84.1 kg

Table S.3
Comparison of Estimated Greenhouse Gas Emission Reductions for News Delivery and Personal Mobility, Per Capita and Total

Factor	Written News	Personal Mobility (maximum adoption case)
Reductions per user		
Current annual GHG emissions (kg CO_2e)	94.7	2,380
Maximum potential annual GHG emissions avoided with alternative service provision (kg CO_2e)	84.3	893
Maximum percentage reduction	89	37
Total reductions		
Current annual GHG emissions for sector (millions of metric tons)	4.35	1,071
Assumed percentage of users who adopt alternative service provision	100	12.5
Maximum potential annual GHG emissions avoided with alternative service provision (kg CO_2e)	3.87	18.1
Maximum percentage reduction	89	1.7

of CO_2e per user, or 3.87 million metric tons overall. Reducing GHG emissions by similar amounts by improving the efficiency of publishing paper newspapers is infeasible. Even in the future, technical improvements in the efficiency of producing paper or printing newspapers could not generate such large reductions in energy use or GHG emissions. Although the total emissions of 4.35 million metric tons of GHG emissions from the newspaper industry are not a large total of U.S. GHG emissions, they represent a substantial reduction for one sector.

Vehicle-sharing programs also generate substantial per-user reductions in GHG emissions, about 37 percent per user. However, only 0.27 percent of U.S. drivers currently participate in vehicle-sharing programs; maximum potential participation has been estimated at 12.5 percent. Participation is constrained by the availability of public transportation, the availability of vehicle-sharing services, and ease of use. Moreover, the profile of those who join vehicle-sharing organizations is quite different from those who do not: Members of vehicle-sharing organizations drive far fewer miles annually than the average U.S. driver. In the unlikely event that maximum potential participation is achieved, the corresponding reduction in total U.S. GHG emissions would be on the order of 18.1 million metric tons. Because potential likely members drive so much less than the national average, this reduction in emissions would be equivalent to only 1.7 percent of total GHG emissions (1,071 million metric tons) from light-duty vehicles. Thus, although the relative size of reductions may be large, overall reductions in the context of total U.S. emissions of GHGs would likely be relatively small.

Acknowledgments

We extend our thanks to Gian Porro and Philip Farese of the National Renewable Energy Laboratory and Darrell Beschen and Seungwook Ma at the U.S. Department of Energy Office of Energy Efficiency and Renewable Energy. They provided valuable insights into the concept of energy services analysis and helped guide the research. Our reviewers, Tom LaTourrette, a senior physical scientist at RAND, and Susan A. Shaheen, Honda Distinguished Scholar in Transportation at the University of California at Davis and codirector of the Transportation Sustainability Research Center at the University of California at Berkeley, made some excellent suggestions, and the report is stronger for them.

We would also like to thank several RAND staff who have made valuable contributions to this report. Johanna Zmud provided insight on transportation data sources. Frederick S. Pardee RAND Graduate School student Xiao Wang assisted with analyzing transportation data from the National Household Travel Survey. Jennifer Miller assembled the manuscript. Christine Galione, formerly of RAND, prepared the reference section.

Abbreviations

AAA	American Automobile Association
Btu	British thermal unit
CASD	computer-assisted scheduling and dispatching
CO_2e	carbon dioxide equivalent
DOE	U.S. Department of Energy
EEED	Environment, Energy, and Economic Development Program
EIA	U.S. Energy Information Administration
EJ	exajoule
EPA	U.S. Environmental Protection Agency
ESA	energy services analysis
GHG	greenhouse gas
HOV	high-occupancy vehicle
ICT	information and communication technology
IMR	Innovative Mobility Research
ISE	RAND Infrastructure, Safety, and Environment
ISO	International Organization for Standardization
LCA	life-cycle assessment
LUV	local-use vehicle
MJ	megajoule (1 million joules)
MSA	metropolitan statistical area
NAA	Newspaper Association of America
NHTS	National Household Travel Survey
NREL	National Renewable Energy Laboratory

R&D	research and development
SAIC	Science Applications International Corporation
TJ	terajoule (1 trillion joules)
VKT	vehicle kilometers traveled
VMT	vehicle miles traveled

CHAPTER ONE

Introduction

Definition of *Energy Services Analysis*

This report employs energy services analysis (ESA) to illustrate how alternative ways of satisfying human wants and needs can reduce energy consumption and associated emissions of greenhouse gases (GHGs). We define *ESA* as an analytical approach that investigates opportunities for reducing energy use by adopting alternative means of satisfying human wants and needs. This focus on satisfying wants and needs contrasts with other approaches that analyze means to reduce energy consumption through technological or behavioral changes but retain existing patterns of consumption of goods or services.

This report illustrates the ESA approach with two case studies: (1) providing print news by electronic dissemination instead of newspapers and (2) providing mobility through vehicle-sharing services instead of through the private ownership of motor vehicles. The first, electronic delivery of daily news, is already commercially available, has shown signs of capturing a substantial share of the market for news, and has the potential to transform news delivery. The second, vehicle sharing as a way to provide personal mobility, is commercially available but has thus far not been transformative.

For each service, we first describe the alternative way in which the service is delivered. We then estimate the potential reduction in energy consumption and associated GHGs that the alternative service might generate, drawing on previous analyses of the energy intensity of the conventional and new forms of delivery. We then developed two models, one for electronic dissemination of news via e-readers and tablet computers and one for vehicle sharing, to estimate the aggregate potential reduction in energy use and GHG emissions in the United States that wider-scale adoption of these technologies might generate. Subsequently, we examined technical, commercial, and behavioral factors that would encourage or impede the adoption of the alternative means of delivering the service.

We utilized existing analytical methodologies to conduct ESA, including process analysis and life-cycle energy analysis. We focused on understanding how energy is currently used to provide the services and how that would change with alternative means of provision.

A Word on Measurement

Throughout this report, we use several measurements to capture energy use and GHG emissions that bear explanation. We generally use joules or megajoules to measure energy because this is the measure used in many scientific journals. However, we use miles per gallon to measure fuel efficiency for motor vehicles. Kilowatt-hours are used to measure the use of electricity. Because of these differences, which make it difficult to compare energy use across sectors, this report focuses on GHG emissions as the key metric.

The standard measure of GHG emissions is in terms of kilograms (kg) of carbon dioxide equivalent (CO_2e). This measure converts all GHGs, such as methane, water vapor, and other gases that trap heat within the atmosphere, into a common measure: the equivalent amount of carbon dioxide that would have the same effect in terms of trapping heat. For larger amounts, we refer to metric tons (1,000 kg equals one metric ton) or million metric tons.

Organization of This Report

Following this introduction, this report is organized into five chapters. Chapter Two introduces ESA in more detail, provides examples of how ESA can be applied to different services, and presents alternatives to traditional approaches to analyzing energy consumption from the perspective of end use. Chapter Three presents our analysis of e-readers and other electronic devices as a different way to deliver news, comparing the GHG emissions of newspapers and e-readers. Chapter Four presents our analysis of vehicle sharing as a new way to provide personal mobility. Chapter Five presents several additional examples of potential ways to deliver services through alternative means that save energy compared with current means of providing a service. In most instances, these alternatives are not yet commercially available but could become so in the not-too-distant future. Chapter Six presents our key conclusions, and an appendix shows the conversion of energy use from the original data to kilograms of CO_2e.

CHAPTER TWO

Energy Services Analysis

Energy services has long been defined as the ultimate uses for which energy is consumed (Reister and Devine, 1981). The demand for energy services is a derived demand: People do not use energy for its own sake but as an input to generate goods or services consumed to satisfy human wants and needs. ESA is therefore a framework to study and ideally reduce energy consumption based on the provision of an end-use service rather than by energy sectors.

Although the concept of ESA has been discussed and applied for the past several decades, the term itself has not entered widespread use. Various papers that use or promote what this paper defines as ESA have instead referred to "end-use energy efficiency" (Jochem et al., 2000), "analysis of energy services" (Hillsman, 1991), "energy needs audit" (Sustainable Energy Authority of Ireland, undated), and analysis of household or residential (Sovacool, 2011) or lifestyle (Weber and Perrels, 2000) energy services. This framework has been applied in a variety of contexts, including end-use industrial energy (Beer, 2000) and household consumption (Weber and Perrels, 2000), and using different categories of wants and needs. The goal is always to seek ways to provide the same amount of service with less energy (Haas et al., 2008).

This chapter compares ESA with conventional energy-use analysis, discusses its strengths and weaknesses, and provides several analytical frameworks for conducting ESA.

Energy Services Analysis and Analytical Approaches That Are More Conventional

One commonly used approach to assess opportunities for reducing consumption of energy focuses on improving end-use efficiency. To assist policymakers and analysts in cataloging and compiling opportunities for reducing energy end use, the U.S. Energy Information Administration (EIA) aggregates end use of energy into four broad consumption categories: residential, commercial, industrial, and transportation. These categories are often used to identify areas in which energy consumption could be reduced or technologies could be made more energy efficient (Bin and Dowlatabadi, 2005).

This conventional approach has two shortcomings. The first is that it focuses on energy consumption per se as opposed to the wants and needs fulfilled by the goods and services that the energy is used to produce. The second is that it often emphasizes direct, as opposed to total, energy use in providing a service.

In response to the first shortcoming, ESA accounts for the fact that energy is often an important input used to generate goods and services that people desire to satisfy their wants and needs. This expands the universe of potential means of reducing energy use from improving the energy efficiency of existing goods and services to new ways to meet these wants and needs that would not be considered in a conventional analysis. By focusing on alternative means of satisfying wants and needs, rather than on improving the efficiency of technologies used to produce goods and services currently consumed to address these wants and needs, ESA uncovers new opportunities and alternatives that could potentially lead to substantial reductions in energy use and GHG emissions. ESA provides analytical insights that approaches focused solely on technologies and their associated energy use may fail to uncover.

One approach to analyzing and thereby focusing on wants and needs is to categorize various activities in accordance with their effects on one's quality of life. Costanza et al. (2007) proposes a definition of *quality of life* that is made up of these wants and needs: subsistence, reproduction, security, affection, understanding, participation, leisure, spirituality, creativity, identity, and freedom.

Costanza et al. assign various activities to these determinants of quality of life. For example, the consumption of food and use of shelter both fulfill the need for subsistence. Watching television and playing table tennis are leisure pursuits. These categories can help focus analysts on activities that address wants and needs rather than on the production of specific goods or services. By focusing on the underlying want, individuals may discover opportunities for reducing energy consumption by satisfying that desire in a different way.

Drawing on the list developed by Costanza et al., Table 2.1 matches human wants and needs (left column) with means by which these wants or needs are satisfied (middle column). The right column shows services that can be provided in part by the use of energy to meet these wants and needs.

This mapping is not perfect. Some cells in the right column are blank because the inputs needed for these as identified in Costanza et al. are mostly human capital and time, inputs that are difficult to link to energy consumption. In other cases, the services are applicable to more than one need. For example, mobility is important not only as a means of directly satisfying the need for freedom but also in transporting food for subsistence and in making it possible for people to visit scenic areas to enjoy nature or houses of worship to participate in religious services.

The second shortcoming of the conventional approach is that it often emphasizes direct, as opposed to total, energy use in providing a service. Total energy use is the sum of direct and indirect energy use. Direct energy use is the energy consumed in the phase of producing a good or service. Indirect energy is all the energy used to make the production of a good or service possible: e.g., manufacturing the machinery to produce the item, transporting components and machinery to the plant site, distributing the product. (Because indirect energy includes these preceding steps, it is sometimes referred to as *embodied energy* [Bin and Dowlatabadi, 2005].) For example, the energy used to wash clothes includes not only the electricity or natural gas used to operate the washer and dryer; it also entails the energy that went into manufacturing the washer and dryer, transporting them to the home, and disposing of them when they come to the end of their service lives. Indirect energy use often accounts for a major share of all the energy uses to produce a good or service. Although this might be captured in a conventional analysis, it is sometimes divided between different end-use categories. In the laundry example, the direct energy is used in the residential sector, but the indirect energy is divided across industrial (to manufacture the washing machine), transportation (from the factory to the store and again to the purchaser and, finally, when it is disposed of), and commercial (the

Table 2.1
Mapping of Human Wants and Needs, Direct Satisfiers, and Services Provided by Energy Use

Human Want or Need	Direct Satisfier	Service Provided by Energy Use
Subsistence	Food, shelter, vital ecological services (e.g., clean air and water) health care, rest	Production and distribution of food and shelter, lighting, heating and cooling, provision of health care
Reproduction	Nurturing children, pregnancy, transmission of culture, homemaking	Provision of health care, provision of house cleaning
Security	Enforced, predictable rules of conduct; safety from violence at home and in public; security of subsistence into the future; maintaining a safe distance from crossing critical ecological thresholds; stewardship of nature to ensure subsistence into the future; care for the sick and elderly	Collection and disposal or reuse of waste, provision of health care
Affection	Solidarity, respect, tolerance, generosity, passion, receptiveness	Mobility (the ability to be with loved ones)
Understanding	Access to information, intuition and rationality	Creation and delivery of news, analysis, and stories
Participation	Acting meaningfully in the world; contributing to and having some control over political, community, and social life; being heard; meaningful employment; citizenship	Mobility (access to employment and community and political events)
Leisure	Recreation, relaxation, tranquility, access to nature, travel	Mobility (access to recreational areas)
Spirituality	Engaging in transcendent experiences, access to nature, participation in a community of faith	Mobility (access to religious institutions)
Creativity or emotional expression	Play, imagination, inventiveness, artistic expression	
Identity	Status, recognition, sense of belonging, differentiation, sense of place	
Freedom	Noninterference with certain choices that are especially personal and definitive of selfhood, mobility	

SOURCE: Costanza et al., 2007, Table 1, with our additions.

retail outlet where the unit was purchased). This breakdown between categories makes it more difficult to analyze the total energy used to wash clothes.

In contrast, ESA explicitly takes into account both the energy used in providing the good or service and the energy embedded in the supporting equipment and infrastructure that make production of the good or service possible. By aggregating data on energy used both directly and as embodied in supporting infrastructure, ESA makes it possible to measure how much energy is used across the entire service. ESA thereby helps analysts and policymakers identify areas in which research and development (R&D) and initiatives to transform the provision of particular services can contribute to reducing energy use and GHG emissions.

Instead of seeking to make existing goods and services more energy efficient, the end result of an ESA effort may well be substitution. For example, people do not demand energy for transportation per se; rather, they use transportation to fulfill a variety of needs, from buying food at a grocery store for subsistence to driving to a club because it fosters social interaction. Identifying transportation as a single service, rather than one that facilitates many needs, may lead researchers and policymakers to overlook opportunities to improve energy efficiency by substituting other activities to achieve the same goal. For example, individuals with common interests could interact through the Internet or on the telephone rather than by driving to meet face to face.

Similar benefits can be derived by considering specific goods or services. For example, Vringer and Blok (2000) estimate the indirect energy requirements of a consumer good that fulfills a want for emotional expression: arranging and enjoying cut flowers. The authors estimate that household decoration is responsible for approximately 7 percent of household energy use in the Netherlands, with cut flowers accounting for more than one-quarter of this requirement. They suggest two changes that could reduce the energy requirement from this service. One, changing flower purchasing patterns, such as buying flowers grown in summertime and in warm countries. Two, recognizing that flowers provide an aesthetic service, that service could be provided by growing household plants, displaying flowers made of textiles, or artwork.

The total amount of energy measured by conventional approaches and ESA may differ. Conventional energy sectors measure energy used to produce goods and services domestically, including energy used to manufacture exports. In contrast, ESA evaluates all the energy consumed in the country to provide goods and services, including the energy used to generate imported goods and services. For example, the United States imports a substantial number of cars. ESA would measure the energy consumed to manufacture and ship the imported cars and ascribe that use to the United States, not to the countries where the cars are manufactured.

ESA has several other benefits and drawbacks as well. It may be able to test assumptions that are often accepted as truisms; for example, the claim that locally produced food requires less energy to transport to market than food grown in distant locations. However, it may also be that energy use is more difficult to analyze because it can require data that may not be readily available. In addition, if it leads to an analysis about substituting one way of providing a service for another, it may be difficult to compare the two because of inherent differences in how services are delivered. For example, in Chapter Three, we compare print and electronic delivery of written news; although one can replace the other, they are not direct substitutes the way comparing one printing process with another would be. Finally, it may be difficult to agree on what service is actually being provided. For example, two vehicle trips may serve very different purposes, so they may fall into two different categories of services with very different implications for substitution, even though the trips themselves are identical.

Although ESA may not be ideally suited for every type of energy use, it does seem to lend itself to several situations. In markets in which new technologies are emerging, it can help determine whether there is an advantage to faster adoption. It may be more useful for services consumed at the household level than at the commercial or industrial level because the household is the "ultimate" consumer, while services used at the commercial or industrial level are only one step in a longer chain.

Conducting Energy Services Analysis

ESA involves several steps. First, the analyst needs to carefully define the want or need around which the analysis is to revolve. Second, the analyst needs to conduct a comprehensive review of means by which the want or need could be satisfied. Third, the analyst needs to define the level of the analysis: the individual, the household, or society in general. Fourth, the analyst needs to select techniques to measure and compare total energy use across the various alternatives for satisfying the need or want.

In the previous section, we discussed one typology from the psychological literature for categorizing wants and needs and identifying means by which those wants and needs are satisfied. Identifying such alternative means may entail interviews with households about what are seen as substitutes for fulfilling various wants and needs. For example, surveys of time use by individuals have identified trade-offs among various activities. Satisfactory substitutes differ depending on the purpose of the activity. For example, some individuals consider time spent exercising as leisure; others, as part of their pursuit of health. Satisfactory substitutes vary quite dramatically depending on which of these purposes are paramount.

ESA may be conducted at different levels of aggregation. Focusing on the individual is helpful for finely differentiating among alternatives. However, for analysts concerned about aggregate levels of energy use and GHG emissions, national aggregate analyses are needed.

In the next section, we discuss some of the standard analytical techniques used in ESA to measure energy.

Tools for Conducting Energy Services Analysis

Process Analysis

A standard industrial technique for measuring energy consumption is process analysis. In process analysis, a manufacturer or supplier identifies each discrete step in the chain of activities needed to produce and deliver a good or service. Analysts then measure the energy consumed at each of these steps. The process analysis method is computationally intensive and requires highly disaggregated data (Park and Heo, 2007).

To assist in conducting process analyses, the World Business Council for Sustainable Development and the World Resources Institute have developed a GHG protocol to help firms quantify GHG emissions from their activities using process analysis (Bhatia and Ranganathan, 2004). The protocol specifies how firms should measure three types of emissions:

- Scope 1 emissions are produced by sources that are owned or controlled by the company.
- Scope 2 emissions are produced from the generation of purchased electricity.
- Scope 3 emissions are "a consequence of the activities of the company, but occur from sources not owned or controlled by the company" (Bhatia and Ranganathan, 2004, p. 25).

Life-Cycle Assessments

Whereas process analysis is generally conducted at the firm level and often bounded by the process of manufacturing or producing a good or service, life-cycle assessments (LCAs) are focused on a single product or service. LCA evaluates energy use over the entire life of the product or service: the energy consumed in the production, distribution, use, maintenance, and disposal of a product or service (Science Applications International Corporation [SAIC], 2005). It is "cradle-to-grave" analysis because it considers all aspects of a product's life cycle, from extraction of raw materials to final disposal. When used to conduct ESA, LCA provides the data needed to compare the energy consumed by alternative means of satisfying a want or need.

The International Organization for Standardization (ISO) has suggested standardized guidelines for the LCA process in its ISO 14040 series. The standard framework for LCA involves four steps:

1. Define the goal and scope of the analysis, including the boundaries of the system to be studied and the environmental impacts to be considered.
2. Identify and quantify inputs of energy, water, and other materials, as well as emissions.
3. Evaluate potential human health and ecological impacts of the input use and emissions identified in the previous step.
4. Evaluate potential options for reducing environmental impacts (SAIC, 2005; Blottnitz and Curran, 2007).

The United Nations Environment Programme and the Society of Environmental Toxicology and Chemistry have developed a consortium of private and public partners interested in performing LCA and have created an inventory of LCA databases available throughout the world. In the United States, the most prominent LCA database is hosted by the National Renewable Energy Laboratory (NREL), which maintains a database of material and energy flows in common units for a large number of industrial processes in the United States (Curran and Notten, 2006; NREL, 2012).

By way of illustration of LCA analyses, we offer the following example. Levi Strauss conducted an LCA of a pair of its 501® jeans. It found that, over the life cycle of these jeans, the jeans were responsible for 32 kg of CO_2e emissions, primarily from growing the cotton used in the jeans and from after-sale washing and drying. The study determined that advances in laundry detergents meant that jeans could be washed in cold water (rather than warm water). This, along with other measures, suggested that product life-cycle GHG emissions could be reduced by as much as 50 percent, primarily by the company changing washing instructions from washing in warm water to cold (Kahn, 2009).

Observations on Applying Energy Services Analysis

Like conventional consumption categories, each of these service categories has limits. The diversity within and between these groups, as well as the difficulty of developing a single approach that can illuminate all energy uses and options, precludes such an all-encompassing analysis.

In the following two chapters, we illustrate the use of ESA through case studies. In each case, we use ESA to evaluate the GHG emissions from a particular service. We estimate the potential GHG emission reductions from moving from a baseline case to an alternative case in which the same service is provided in a different manner. We then discuss the feasibility of widespread adoption of the alternative case and suggest potential measures that the U.S. Department of Energy (DOE) could take to encourage adoption of the alternative.

CHAPTER THREE

Communications: Electronic Delivery of Daily Written News

In this chapter, we use ESA to analyze possibilities for reducing energy consumption and GHG emissions by delivering the written daily news through means other than newspapers. Using a conventional approach, we would investigate this problem by exploring potential technological solutions for reducing the energy used to print and distribute newspapers. The exploration would focus on reducing energy consumption in harvesting pulpwood, making paper, printing, and distribution. Using ESA, we explore an alternative means of providing written news: electronic delivery. To investigate potential reductions in energy use and GHG emissions, we compare this alternative approach of delivering news with providing news through newspapers. Using ESA, the focus shifts from improving the energy efficiency by which newspapers are currently produced to assessing energy savings if consumers are able to satisfy their desire for written news using alternative means.

To measure potential savings in energy use and reductions in GHG emissions, we compare the energy used and GHGs emitted over the entire span of printing, distributing, and disposing of newspapers with those from electronic delivery of news, including the manufacture, operation, and disposal of two electronic devices: tablet computers and e-readers. We use process and life-cycle analyses conducted by other researchers to make these calculations. We then extrapolate from energy savings for a single subscriber to estimate potential aggregate reductions in energy use and GHG emissions for the United States as a whole under various assumptions about the level of substitution of e-readers for newspapers.

We find that one newspaper subscription emits an average of 94.7 kg of CO_2e per year (see Table 3.1 later in this chapter). If we assume that one paper subscription is replaced with an e-reader or tablet computer, that replacement results in a net reduction of GHG emissions of 63 to 74 percent annually, or 2.7 million to 3.2 million metric tons annually, if the device were used solely to read newspapers.

Energy Consumption and Greenhouse Gas Emissions from Newspapers

Delivery of written news is one way to meet the demand for access to information and, in turn, the human desire for understanding. Although news is distributed over radio, television, and other media, written news retains some advantages over television and radio in that newspaper readers are able to absorb the news at their own pace, and they can reread or save articles. Readers also find it easier to select news items that are of greatest interest to them than when listening to the radio or watching television.

Printing and distributing newspapers entails several steps, some of which use substantial amounts of energy. Newsprint is manufactured from wood pulp, other plant materials, or recycled paper and transported to printing plants. Formatted files are sent to printing plants; from there, newspapers are shipped to newsstands or delivered directly to subscribers.

Previous Studies Estimating Energy Use and Greenhouse Gas Emissions from Newspapers

Several studies have estimated energy use and GHG emissions from printing newspapers and other printed materials. These studies have used a variety of methodologies, most commonly process analysis, and set different system boundaries for their analyses. Some of these efforts have been conducted by small, in-house teams, while others have relied on consultants or teams from several organizations that span supply chains. Some have focused on newspaper production only, while others have examined distribution as well. Some efforts have been industry-wide; others have focused on one or two publications. Finally, some have looked broadly across the supply chain to understand the interactions between different types of products. Following is a list of some key studies:

- Barone et al. (2008) present an effort by *DISCOVER Magazine* staff to work with fellow employees and suppliers to estimate the GHG emissions from one issue of the magazine.
- Binns et al. (2008) present the results of a Markets Initiative and Green Press Initiative effort to assess the environmental impact of the U.S. newspaper industry and to identify opportunities for improvement.
- Borealis Centre for Environment and Trade Research (2008) summarizes the results of an effort sponsored by the Book Industry Study Group and Green Press Initiative to use surveys to estimate GHG emissions from the book publishing industry.
- Borkowski and Kelley (2001) summarize the efforts of three environmental nonprofit organizations to estimate the environmental impact of the magazine industry and to offer recommendations for improving environmental performance.
- Carbon Trust (2006) did a pilot study of emissions from a daily newspaper and weekly magazine published by Trinity Media, the UK's largest newspaper publisher.
- Gower et al. (2006) summarize the efforts of the H. John Heinz III Center for Science, Economics, and the Environment and several partners to understand GHG emissions associated with growing and harvesting trees in Canada (by Canfor Corporation), the conversion of those trees to lumber (sold in Home Depot stores) and wood pulp (used by Stora Enso North America) in the United States, and the production and distribution of *Time* and *InStyle* magazines (published by Time).
- Kinsella et al. (2007) summarize a collaborative effort of ten environmental nonprofit organizations to estimate the environmental impact of the paper industry.
- Nors, Pajula, and Pihkola (2009) analyze the carbon footprint of a typical Finnish daily newspaper and weekly magazine.
- Pihkola et al. (2010) conduct a detailed analysis of emissions of conventional pollutants and GHG from a fictitious Finnish daily newspaper.

Estimating Greenhouse Gas Emissions from Newspapers

For the purpose of this analysis, we have drawn on the study of Pihkola et al. (2010) for GHG emissions from the production of a fictitious but typical Finnish daily newspaper. Of the studies we reviewed, it was the most comprehensive analysis of a newspaper's production and the associated emissions. Table 3.1 shows the data and categories we have used to estimate GHG emissions released by printing and distributing one newspaper.

Using the Finnish figures, we developed a similar set of estimates for the United States, with the only difference being the mix of purchased energy (electricity generation). Finland has a higher percentage of hydropower, nuclear, and renewables in its generating mix than the United States has. Consequently, the United States emits 54 percent more GHG than does Finland per kilowatt-hour (Carbon Trust, 2006, Figure 13). We have incorporated this difference in emissions into the analysis by adjusting emissions from purchased energy for our estimates of GHG emissions per newspaper in the United States.

From this analysis, we find that, in the United States, one newspaper with an assumed weight of 0.2 kg results in the emissions of 0.254 kg of GHGs. This compares with 0.213 kg for the same newspaper in Finland. (For this analysis, we exclude GHG emissions from the work of researching and writing news stories because we argue that this will be the same regardless of whether the story appears in print or online.)

For sake of comparison, we contrast these estimates with those from a similar analysis of GHG emissions per issue of the *Daily Mirror*, a UK newspaper published by Trinity Mirror,

Table 3.1
Greenhouse Gas Emissions from Printing and Distributing a Newspaper

Factor	Finland	United States
Weight of prototypical newspaper (kg)	0.2	0.2
Production step (kg of CO_2/1,000 kg of newspaper)		
Paper production	565.0	744.4
Fiber supply	10.7	10.7
Chemicals, materials, fuels	42.6	42.6
Direct emissions	181.2	181.2
Emissions from purchased energy	330.5	509.9
Printing	138.6	190.7
Chemicals, materials, fuels	42.6	42.6
Emissions from purchased energy	95.9	148.0
Delivery	170.6	170.6
Other transport	21.3	21.3
End of life	170.6	170.6
Total GHG emissions per metric ton of newspapers (kg of CO_2/1,000 kg of newspaper)	1,066.0	1,297.6
GHG emissions per issue of newspaper (kg of CO_2/1,000 kg of newspaper)	0.213	0.254
GHG emissions per newspaper subscription, assuming 365 papers per year (kg of CO_2/1,000 kg of newspaper)	77.8	94.7

SOURCE: Pihkola et al., 2010, Figure 14.

NOTE: We multiplied the percentages in Pihkola et al. (2010, Figure 14) by Pihkola et al.'s total figure for GHG emissions per issue of newspaper, 1,066 kg. We calculated U.S. figures for emissions from purchased energy based on information provided by Carbon Trust (2006, Figure 13).

a UK company. That study found that each newspaper (average weight 0.182 kg) resulted in 0.174 kg of GHG emissions (Carbon Trust, 2006, p. 15). To make the two studies comparable, we have standardized the Carbon Trust figures on the heavier newspaper assumed by Pihkola et al. Assuming that the *Daily Mirror* averaged 0.2 kg in weight, each issue would result in GHG emissions of 0.191 kg, compared with 0.213 kg in the Finnish study; the Finnish estimates are 12 percent higher than the Carbon Trust's.

Differences in three assumptions could drastically change Pihkola et al.'s results, our results reported in Table 3.1, and the results from the Carbon Trust study. The first assumption is the share of recycled paper in the paper source. In their analysis, Pihkola et al. assume that recycled paper accounts for 60 percent of the inputs needed for newsprint, with pulp and fillers accounting for 35 and 5 percent, respectively. Using parameters from the Carbon Trust (2006) study, we estimate that GHG emissions would decline by about 115 kg per metric ton of newsprint, or 9 percent of the emissions associated with producing 1 metric ton of U.S. newspapers, if the share of recycled paper in newsprint were increased from 60 percent to 100 percent.[1]

The second assumption is the source of purchased energy for electricity. We calculate that, if electricity generated by renewable sources of energy or nuclear energy replaced the current U.S. fuel mix, GHG emissions would decline by close to 658 kg per metric ton of newsprint or by close to half of total GHG emissions.

The third assumption is how consumers dispose of newspapers. In the United States, almost three-quarters of newspapers are recycled (Newspaper Association of America [NAA], undated). In the Finnish analysis, 79 percent of papers are assumed to be recycled. If 100 percent of newspapers are recycled, emissions would decline by 66 to 96 kg per metric ton of newsprint (Pihkola et al., 2010, p. 75), or 5 to 7 percent compared with our estimate of total GHG emissions in the U.S. context. The decline is due to the elimination of emissions of methane and carbon dioxide associated with the decomposition of paper in landfills.

Energy Use and Greenhouse Gas Emissions Associated with E-Readers and Tablet Computers

In the past few decades, consumers have gradually been shifting from reading newspapers to obtaining written news electronically (Perez-Pena, 2009; Ovide, 2009; Pew Research Center for the People and the Press, 2009). This shift has accelerated in the past two decades because of the spread of online information through the Internet and the diffusion of personal computers, tablet computers, and e-readers.

In this section, we assess energy consumption and GHG emissions from two means of receiving and reading news electronically: e-readers and tablet computers. We focus on these two electronic devices because the former is exclusively dedicated to reading and the latter is the most similar in terms of usage among electronic devices with more capabilities. Their backlit screens have the look of conventional printed material, eliminating glare and other unpleasant aspects of reading on a screen. Font size can be adjusted to make pages more readable for

---

[1]   Carbon Trust (2006) states that raising the share of recycled paper used in newsprint from 50 percent to 100 percent results in a 27 percent decline in energy consumption. Linearly extrapolating this figure and applying it to our estimates of emissions of purchased energy used in paper production, we generate the figure of 115 kg per metric ton of newsprint cited above.

those with poor eyesight. Content can be accessed online. E-readers are also considerably more energy efficient than other portable devices, allowing them to be used for longer periods before recharging.

Tablet computers are substantially more versatile than e-readers, making possible a large range of applications, including most of the functions of desktop and laptop computers. In addition to reading, users can use tablets to view television shows and films, listen to music, and enjoy other forms of entertainment. Newer versions of tablet computers use screens that are more readable and more energy efficient, providing users with longer periods between charging.

Previous Studies Estimating Energy Use and Greenhouse Gas Emissions from Tablet Computers and E-Readers

A few studies have assessed energy use and GHG emissions from e-readers and tablet computers. In one early study, Kozak (2003) analyzed the energy consumed in reading college textbooks on the RCA RAB 1100, an e-reader introduced in 2001. Moberg et al. (2009) analyzed the energy and associated GHG emissions from paper newspapers, reading the news on the Internet, and reading it on an e-reader, the iRex Iliad, that was introduced in 2006. Each of these studies measured the energy used to manufacture and dispose of the devices, operate them, and run the servers and communication equipment used to store and transfer information to them.

However, these studies did not have access to proprietary data on the indirect energy used to manufacture the devices and their proprietary components. Therefore, they estimated the energy used to manufacture these devices employing a variety of techniques. Moberg et al. (2009) identified the raw materials used to produce the devices, estimating the energy required to produce those materials, and then calculated the energy required to manufacture at least some of the components. Kozak (2003) scaled previous studies on the energy required to manufacture key components, including microprocessors, memory chips, and liquid crystal displays. However, some components were excluded because of a lack of data.

Each study adopts its own approach to measuring energy use and GHG emissions. The two studies differ in terms of whether results are measured on the basis of GHG emissions or energy used and whether energy use is measured on an end-use or primary energy basis.[2] Each also adopts different assumptions for systemwide efficiency of the electric power system, the GHG intensity of electricity generation, e-reader life spans, and e-reader usage.[3] In our analysis, we have standardized approaches across the devices so that the estimates are fully comparable.

---

[2] End-use energy is a measure of the energy used at the point of consumption (e.g., electricity consumption), whereas primary energy use represents the amount of fossil, nuclear, or renewable energy needed to generate power for end use. For example, using U.S. national averages, nearly three times the amount of primary energy is consumed to generate, to transmit, and for electricity than the energy embodied in a kilowatt-hour delivered to a consumer. The difference is the result of the efficiency losses caused by converting one form of energy to electrical energy in a power plant and then transmitting and distributing that electricity through the power grid. In 2008 in the United States, 40.67 quadrillion British thermal units (Btu) were consumed to generate electricity, but only 13.21 quadrillion Btu were delivered for end use, yielding an average U.S. electric power system efficiency of 32.5 percent (EIA, 2009, p. 225).

[3] Kozak (2003) uses systemwide efficiency information and GHG emissions for electric power generation in certain states in the Midwestern United States and California (in-state generation only); Moberg et al. (2009) use data from Europe and China (for manufacturing).

Estimating Greenhouse Gas Emissions from Tablet Computers and E-Readers

Because of manufacturers' investments in technological development due to rapidly growing demand, both tablet computers and e-readers are changing quickly, meaning that the assumptions and figures we use in the analysis may become outdated quickly. Manufacturers are adding new features while extending battery life and reducing energy use. For example, Apple reduced energy use and GHG emissions by 19 percent between the introductions of the iPad and the iPad 2 while adding many new capabilities (Apple, 2010, 2011). In light of these rapid changes in the marketplace, it is unclear whether consumers will gravitate to one technology or the other to receive written news.

The two technologies continue to differ in terms of energy use and associated GHG emissions. Because of the display technology, tablet computers require more energy to operate than e-readers. To capture these differences and the uncertainties of the marketplace, we provide estimates of associated GHG emissions for both types of devices, using the latest estimates that we were able to find on energy consumption or GHG emissions. For tablet computers, we use information that Apple publishes in its environmental report on the iPad 2 (Apple, 2011) on life-cycle GHG emissions for that device. The report provides limited methodological details but is calculated in accordance with the guidelines and requirements specified by ISO 14040 and ISO 14044.

For e-readers, we have drawn upon the Moberg et al. (2009) analysis of a Swedish e-reader, the iRex Iliad. For their analysis of GHG emissions from receiving a newspaper by e-reader, Moberg et al. (2009) use Swedish and European data on the mix of fuels used to generate electricity. Because Sweden uses much less coal than the United States in its generating mix, Moberg et al.'s estimates of GHG emissions from e-readers, if applied to the United States, would be biased downward. In our analysis, we calculate likely GHG emissions from e-readers in the United States using Moberg et al.'s data on energy consumption, but we estimate associated GHG emissions based on data on the generating mix and associated GHG emissions in the United States. We use the Iliad because, at the time of this writing, no emission information had been made publicly available by the producers of the three most popular e-readers in the current market: Amazon (Kindle), Barnes and Noble (Nook), or Sony (Reader) (Abell, 2010).

For each technology, we provide estimates of indirect energy use and associated GHG emissions related to manufacturing, distributing, and disposing of the device and direct energy use and associated GHG emissions from charging and using the device and with the information and communication technology (ICT) network from which the written news reports are drawn. In the case of manufacturing, we include the energy and associated GHG emissions used to extract the raw materials, manufacture the components, and assemble and package the product. Energy and associated GHG emissions used in distribution include the energy used to transport the assembled device from the point of production to the final customer. We also measure the energy and associated GHG emissions required to collect, disassemble, recycle, and dispose of the device once it reaches the end of its useful life, which we assume to be three years (Apple, 2011).[4]

---

[4] Moberg et al. (2009) assume that consumers will use an e-reader for one year and then replace it with another, more modern version. This assumption seems more appropriate for an early-stage device than for the wider use of e-readers that exists in 2011. According to the manufacturer of the device analyzed by Moberg et al., the technical life of the reader is four years (Moberg et al., 2009, p. 35). To create a common basis for comparison, we have chosen to adopt Apple's assumption

To measure direct energy use, we include energy used to charge or operate the device. We also include energy used by the device's charger when the device is not connected or when the device is fully charged and the charge is being maintained.

Table 3.2 shows annual GHG emissions for the iPad 2 and the iRex Iliad. As noted earlier, we assume that both devices are used for three years and then discarded. Consequently, indirect energy use was calculated per year by dividing total indirect energy use by three.

Comparing Greenhouse Gas Emissions from Printed Newspapers and Electronically Disseminated Printed News

To what extent could substituting electronically distributed newspapers for paper newspapers contribute to reducing GHG emissions? Table 3.3 compares total emissions from newspapers, tablet computers, and e-readers over the course of one year. It also shows the potential declines in emissions if all paper newspapers were to be displaced by tablet computers or e-readers, ascribing all the energy usage and associated GHG emissions of these devices to reading newspapers.[5] We calculated the potential reduction in emissions using the total number of newspaper subscriptions in the United States in 2009, 45.9 million, the latest year for which data on total weighted (Sunday and weekday) subscriptions are available (NAA, undated). The differ-

Table 3.2
Annual Standardized Comparison of Greenhouse Gas Emissions for a Tablet Computer and E-Reader, Assuming a Three-Year Product Life

Feature	Tablet Computer	E-Reader
Device	iPad 2	iRex Iliad
Year of introduction	2011	2006
Emissions from indirect energy use (kg of CO_2e)		
Manufacturing	21	11.0
Distribution	3.5	0.04
Disposal	0.4	−0.06
Subtotal	24.9	11.0
From device and ICT network energy use	10.1	13.7
Total	35	24.7

SOURCES: Apple (2011); our calculations based on energy use from Moberg et al. (2009; see Table A.1 in the appendix for information on energy use and conversion factors into GHG emissions).

---

of three years. This period is based on Apple's analysis of the market for tablet computers, which we deem similar enough to e-readers in this regard to warrant the same assumption about product lifetime.

[5]   For the sake of simplicity, we assume here that one tablet computer or e-reader replaces one newspaper subscription. This will not necessarily be the case; some households may replace one paper subscription with two tablets or e-readers. However, lacking data on the number of households in which a newspaper read by more than one person is replaced by multiple electronic devices, we have chosen to make a simplifying assumption.

Table 3.3
Annual Greenhouse Gas Emissions Associated with Reading a Newspaper and Reading the News on a Tablet Computer or E-Reader

Factor	Unit per Year	Newspaper (2009)	Tablet Computer	E-Reader
GHG emissions per subscription	Kg of CO_2e	94.7	35	24.7
Reduction compared with newspapers, per subscription	Millions of metric tons		59.7	70
	Percentage		63	74
Newspaper subscriptions	Millions	45.9		
GHG emissions	Millions of metric tons	4.35	1.61	1.13
Reduction compared with newspapers, total	Millions of metric tons		2.74	3.21
	Percentage		63	74

SOURCE: NAA (undated).

ences in emissions between newspapers and these electronic devices are sizable, running from 63 to 74 percent, or 2.7 million to 3.2 million metric tons annually.

Table 3.3 ascribes all the GHG emissions stemming from owning and using these devices to reading newspapers. This assumption is too extreme. Tablet computers and e-readers are used for activities other than reading newspapers. Users of tablet computers watch movies and exchange email on their devices. E-readers are used to read books and magazines in addition to newspapers. To provide a more accurate comparison of GHG emissions associated with reading newspapers in paper form or electronically, we have distributed the GHG emissions associated with owning and operating tablet computers and e-readers across three activities: reading newspapers, books, and magazines.

The *Statistical Abstract of the United States* (U.S. Census Bureau, 2011) reports that, in 2008, Americans spent an average of 169 hours reading newspapers, 128 hours reading magazines, and 104 hours reading leisure books. In other words, newspapers accounted for 42 percent of the total number of hours Americans spent reading; magazines and leisure books accounted for 32 and 26 percent, respectively. We use this share (42 percent) to allocate the share of GHG emissions associated with e-readers to the activity of reading written news.

For the sake of comparison, we apply this same share to tablet computers, despite the many more activities for which tablet computers are used. Table 3.4 shows these results. As can be seen, receiving and reading news on an e-reader or tablet computer generates far fewer GHG emissions than a subscription to a paper newspaper. Under this accounting of GHG emissions from e-readers and tablet computers, the reduction in GHG emissions from an e-reader could be as much as 89 percent. If all newspapers were delivered electronically, total GHG emissions from disseminating written news could be 3.7 million to 3.9 million metric tons less than if distributed through newspapers.

Table 3.4
Annual Greenhouse Gas Emissions Associated with Reading a Newspaper and Reading the News on a Tablet Computer or E-Reader, Assuming That 42 Percent of the Time Spent on the Device Is Spent Reading Newspapers

Factor	Unit per Year	Newspaper (2009)	Tablet Computer	E-Reader
GHG emissions per subscription	Kg of CO_2e	94.7	14.7	10.4
Reduction compared with newspapers, per subscription	Millions of metric tons		80	84.3
	Percentage		84	89
Newspaper subscriptions	Millions	45.9		
GHG emissions	Millions of metric tons	4.35	0.68	0.48
Reduction compared with newspapers, total	Millions of metric tons		3.67	3.87
	Percentage		84	89

SOURCES: NAA (undated); percentage of time spent on e-reader or tablet computer based on data on average hours spent reading newspapers, magazines, and books in 2008 (U.S. Census Bureau, 2011).

Policy Options

As shown in the previous section, substitution of electronic media for newspapers under our assumptions would lead to substantial reductions in GHG emissions. The magnitude of these reductions is determined by the extent to which electronic media displace newspapers, the amount of energy used by e-readers and ICT network equipment, and the amount of energy used to produce, publish, and distribute newsprint.

The choice of whether to receive written news on paper or electronically is driven by markets. For a variety of reasons, including availability of newspapers online, printed newspapers have suffered a sharp decline in readership. Weighted average daily and Sunday paid newspaper circulation fell 22.1 percent from its peak of 62.5 million in 1989 before the advent of the Internet to 45.7 million in 2009; subscriptions have continued to fall (NAA, undated). Some newspapers have responded to circulation declines and the corresponding need to reduce costs by curtailing their delivery areas, reducing the frequency of printing, or ceasing to print paper editions. Others have launched paid electronic versions that either complement or substitute for print newspapers.[6]

Government policies could accelerate this change. In this section are some potential policy options that the U.S. government could adopt if priority were to be given to reducing GHG emissions through the substitution of electronic media for paper.

Provide Consumers with the Right to Retain Access to Information They Have Purchased

The government may wish to work with stakeholders to develop standards for digital property rights so that consumers can be assured that they will have access to information they pur-

---

6 The *Wall Street Journal* offers both Internet-only and joint paper and electronic subscriptions. For the six-month period ending March 31, 2010, the *Wall Street Journal* had 414,000 electronic subscribers out of a total U.S. circulation of 2.1 million (though some, perhaps many, of these electronic subscribers also receive a printed paper) (Plambeck, 2010).

chase, legally acquire, or create, beyond the life of the e-reader or electronic media platform on which the information was delivered.

Encourage Manufacturers to Provide Information on the Energy Efficiency of E-Readers and Information and Communication Technology Equipment

DOE and the U.S. Environmental Protection Agency (EPA) could encourage manufacturers to clearly post information on energy consumption by e-readers and other electronic devices. DOE and EPA could improve consumer awareness of energy performance of e-readers by including them in the ENERGY STAR program. Inclusion would help inform customers about which products are more energy efficient, giving manufacturers an additional incentive to improve energy efficiency.

As part of this effort, DOE and EPA could support the development of a standard, more comprehensive approach to estimating life-cycle energy consumption and GHG emissions for electronic devices, including e-readers and tablet computers, permitting consumers to make decisions that are more informed and providing manufacturers with a level playing field on which to compare their performance.

Monitor Technological Developments in Information and Communication Technology Network Equipment and Usage

As Internet and communication technology continue to evolve, demands on ICT network equipment will increase. For example, rising demand for cloud computing may lead to significant increases in demand for ICT network services. DOE may wish to more closely monitor these trends, looking for opportunities to improve both the measurement and energy efficiency of the ICT network.

CHAPTER FOUR
Personal Transportation: Sharing, Rather Than Owning, Vehicles

Since the introduction of affordable, mass-produced automobiles, cars have provided Americans with a form of mobility that has transformed where people live and work and how they go about their daily lives. The mobility that automobiles provide is, like energy, a derived demand, driven by the wide variety of wants and needs it helps satisfy. As modern societies are currently structured, mobility is needed to buy groceries (subsistence), attend school and religious institutions (understanding, spirituality, participation), or take a vacation (leisure), to use some of the categories listed by Costanza et al. (2007).

In the United States, mobility is provided primarily by personally owned vehicles. In 2009, Americans traveled 3.4 trillion vehicle miles on the nation's roadways; 94 percent of these miles were traveled in a car or truck (National Household Travel Survey [NHTS], undated). Of these miles driven, 28 percent were to and from work, 24 percent were social and recreational, 17 percent were for personal and family errands, and 15 percent for shopping. The remaining 17 percent involved work-related business, school and church, and other such purposes (Santos et al., 2011, Table 6).

In some cases, the services rendered by these trips could have been provided through other means. For example, online courses can be substituted for in-classroom instruction, or Internet chat rooms can replace face-to-face meetings. Some transportation planners have focused on accessibility rather than mobility, arguing that placing common destinations (such as stores and offices) within neighborhoods makes it easier to fulfill many daily needs by walking rather than driving. Alternatively, mobility can be provided by transportation modes other than personally owned cars, such as mass transit and bicycling, which require less energy and produce fewer GHG emissions. However, satisfying many of the needs articulated by Costanza et al. (2007) will continue to entail mobility via automobiles.

In this chapter, we focus on exploring the potential reductions in GHG emissions that would become possible if a relatively new alternative to personally owned motor vehicles for providing mobility were to become more widely adopted: the use of shared vehicles. Vehicle sharing allows members to rent vehicles on a short-term basis, generally for a flat hourly rate, rather than owning and operating their own personal vehicles. Currently, almost all U.S. driving takes place in personally owned vehicles. However, in some cities, vehicle sharing has become an option.

In terms of our ESA framework, vehicle sharing provides an alternative to personally owning motor vehicles for giving people the mobility to pursue their wants and needs, be those shopping to purchase food, engaging in recreation, or visiting friends and relatives. Although detailed data on vehicle miles traveled (VMT) for different trip purposes in shared vehicles are not readily available, it is clear that, in some neighborhoods and circumstances, vehicle sharing could provide the mobility needed to satisfy many wants and needs.

This case study assesses the potential for vehicle-sharing services to provide mobility in ways other than personal ownership of motor vehicles that will reduce energy use and GHG emissions. After defining *vehicle sharing* more precisely, we discuss the mechanisms by which vehicle sharing can reduce emissions and estimate those mechanisms' per-member effects. Then we discuss the current state of vehicle sharing in the United States and some incentives and barriers to adoption. Finally, we develop three potential cases of vehicle-sharing adoption and show how much GHG emissions could be reduced under each case. These cases are based on the dominant model of vehicle sharing, not on emerging options, such as peer-to-peer vehicle sharing.

We conclude by presenting some alternative ideas for how vehicle sharing or similar innovative ideas about ride sharing could become more of a "game changer." Because these ideas are highly speculative, with most ideas in pilot-testing phases, we have not attempted to quantify the effects that such a shift could have on consumption of gasoline and diesel and on GHG emissions. These ideas are presented to provide DOE with insights about how R&D or other policy interventions could influence the adoption of this or other alternative forms of transportation service.

We find that reductions per participant are on the order of 893 kg of CO_2e annually. Compared with personally owning and operating a vehicle, vehicle sharing would reduce GHG emissions on the order of 37 percent per vehicle-sharing member, bearing in mind that vehicle-sharing members drive less than typical drivers. At its maximum potential, vehicle sharing could reduce GHG emissions from the light-duty vehicle sector by 18.1 million metric tons.

Vehicle-Sharing Services

Businesses that rent vehicles have existed for decades, but vehicle sharing differs from the traditional rental business in some important ways. Vehicle-sharing services do the following that traditional rental businesses do not do:

- Provide hourly rates, making it less expensive to use vehicles for short trips than from traditional rental car firms, which usually charge by the day.
- Position vehicles throughout an area so that members can walk to vehicles located in their neighborhoods.
- Use information technology so that users can quickly locate and reserve available vehicles, obtain electronic access to vehicles (using specially issued cards or cell phones), and streamline billing for vehicle usage.
- Provide members with convenient access to many vehicle types so members can select the type of vehicle that best meets their needs for a specific trip (e.g., a compact car for a short errand, a sport utility vehicle for a family trip, or a pickup truck for moving furniture). This makes it easy to use larger vehicles only for trips for which they are needed, and smaller vehicles for other trips.

Vehicle-sharing services in the United States have been run by five types of operators: for-profit, nonprofit, cooperatives, transit systems, and university research programs

(Shaheen, Cohen, and Chung, 2009). The for-profit models have been independent entities; however, auto manufacturers and rental car companies have been entering the market recently (Shaheen, Cohen, and Chung, 2009; Massey, 2010). As of July 2011, the United States had 26 vehicle-sharing organizations with about 560,000 members and 10,000 vehicles (Innovative Mobility Research [IMR], 2011).[1] Currently, the largest player in the U.S. vehicle-sharing market is Zipcar, which globally owns and operates more than 8,000 vehicles and has 575,000 members (Zipcar, 2011).[2] Zipcar acquired Flexcar in 2007, eliminating its main national competitor. Zipcar offers occasional- and frequent-driver plans, with the frequent-driver plans available at four pricing levels based on expected usage.

The landscape for vehicle sharing is changing rapidly; the description above is of the most common current model. In the past several years, some variations have emerged. The most notable are one-way vehicle sharing, peer-to-peer vehicle sharing, and "unattended access" for conventional rental vehicles. The assumptions for our analysis are based on the current model, while these newer variations—none of which is widespread at the time of this writing—are described in the "Policy Options" section at the end of this chapter.

Potential Reductions in Energy Use and Greenhouse Gas Emissions from Vehicle Sharing

Previous Research on Vehicle Sharing and Reductions in Greenhouse Gas Emissions

Two reports have estimated potential reductions in GHG emissions from vehicle sharing. Martin and Shaheen (2011, p. 1074) find that

> The mean observed impact is –0.58 t [metric tons] GHG/year per household, whereas the mean full impact is –0.84 t GHG/year per household.

Martin and Shaheen focus on households because the use of a shared vehicle affects travel decisions for the entire household. The observed impact is the actual increases and decreases in emissions based on actual household driving; this was calculated at 27 percent of previous emissions. The full impact takes into account miles that would have been driven in a personally owned vehicle but were not driven because those miles were forgone or satisfied by a mode other than driving. The full impact is 56 percent (Martin and Shaheen, 2011).

Haefeli et al. (2006, p. XVIII) employ surveys of vehicle-sharing members in Switzerland to determine reductions in GHG emissions accruing from vehicle sharing in 2005:

> [All together], the result for total impact is energy savings in the magnitude of approximately 78.4 TJ (terajoules: one trillion joules) per year. This amount of energy is equivalent to approximately 2.5 million liters of gasoline (660,430 U.S. gallons). Taking the total impact of aggregated energy savings, this calculates out to approximately 1,400 MJ (megajoules: one million joules; 1,400 MJ corresponds to 190 kg of CO_2 emissions) per customer at the end of the year (published in the respective annual report) and 1,450 MJ (corresponds

---

[1] The count provided on the IMR website was 560,572 members and 10,019 vehicles. These figures are based on numbers provided by vehicle-sharing organizations every six months (Shaheen, 2011). Because the actual number likely changes frequently, we have used 560,000 as the approximate number of U.S. members.

[2] Zipcar also has members in Canada and England.

to 200 kg CO_2 emissions) per Car-Sharing customer. When the overall savings for 2005 are calculated per active customers only, the resulting value is approximately 2,100 MJ per customer (corresponds to reduced CO_2 emissions of 290 kg).

Comparing the results of the two studies, Martin and Shaheen find that U.S. households that participated in vehicle sharing reduced GHG emissions by 0.58 metric tons in observed impacts and 0.84 metric tons when all of vehicle sharing's effects on GHG emissions were fully factored in. Haefeli et al. find that active members of vehicle-sharing programs reduced GHG emissions by 0.29 metric tons. The differences in results may not be as large as they seem: The average fuel efficiency of the Swiss motor vehicle fleet is likely greater than the average fuel efficiency of the U.S. fleet. Consequently, opportunities for saving fuel and reducing GHG emissions should be greater when less efficient U.S. personal vehicles are replaced with more-efficient shared vehicles through motor vehicle sharing than in Switzerland, where, on average, personal motor vehicles are more efficient than in the United States. In addition, many U.S. vehicle-sharing members relied more heavily on privately owned vehicles for travel than Europeans. Thus, the modal shift is likely to be higher among U.S. vehicle-sharing members.

In addition to reducing GHG emissions, vehicle sharing also tends to decrease ownership of vehicles among members. A 2008 survey in North America finds that between nine and 13 vehicles are removed from the roads for every shared vehicle and that, among participants, 25 percent sold a vehicle and another 25 percent did not purchase a vehicle when they otherwise would have (Martin and Shaheen, 2010). Another study found that, on average, per shared vehicle, 14.9 participants give up personally owning vehicles (Millard-Ball et al., 2005). Another survey found that average vehicle ownership per household fell from 0.47 to 0.24 vehicles after adopting vehicle sharing (Martin, Shaheen, and Lidicker, 2010). Because average vehicle ownership per household is 1.87 in the United States (Davis, Diegel, and Boundy, 2011), this statistic suggests that those households that adopt vehicle sharing already own far fewer vehicles than the average.

Reducing Energy Use and Greenhouse Gas Emissions Through Vehicle Sharing

We analyze three mechanisms by which providing mobility through vehicle sharing as opposed to personally owned motor vehicles reduces GHG emissions:

- reductions in the number of miles driven by households
- the use of more-efficient vehicles
- reductions in the number of vehicles needed in the total fleet.

Summing these estimates, we developed an estimate of the total reductions in GHG emissions for each driver who moves from personally owning a vehicle to vehicle sharing.

Reductions in the Number of Miles Driven. Research has shown that an important influence on decisions regarding alternative transportation modes for a particular trip is out-of-pocket costs. Because most vehicle trips are made in personally owned vehicles, such that many of the costs of the vehicle are fixed (e.g., the purchase price, insurance), drivers often pay attention only to out-of-pocket costs, such as fuel, tolls, and parking, when deciding whether to drive or take some other transportation mode; they do not factor in the costs associated with owning the vehicle because these costs are incurred whether or not the vehicle is driven for a particular trip. Consequently, driving appears cheaper than other transport modes for each trip because these fixed costs do not go into the decision on transport mode (Steininger, Vogl, and Zettl, 1996).

When individuals participate in vehicle-sharing programs, more of these fixed costs are incorporated into the costs paid per trip. The out-of-pocket costs to individuals for each trip are higher. Not surprisingly, members of vehicle-sharing programs tend to travel less on average than they did when they owned vehicles. One review of nine U.S. vehicle-sharing programs found that, after a member replaces personal ownership with membership in a vehicle-sharing program, his or her VMT falls 7 to 79 percent (although, in some cases, the results were not considered statistically significant) (Shaheen, Cohen, and Chung, 2009). In nine European programs, average per-member VMT fell between 26 and 72 percent (Millard-Ball et al., 2005). Although VMT generally rose among those who had not previously owned a vehicle, those increases were more than canceled out by the reductions in VMT among those who previously owned vehicles (Millard-Ball et al., 2005; Martin and Shaheen, 2010). Using a survey of 9,635 North American vehicle-sharing members conducted in 2008, Martin and Shaheen (2011) found that, on average, VMT fell 27 percent per household; we use this figure as the basis of our calculations.[3]

To calculate potential energy savings and reductions in GHG emissions from abandoning use of a personally owned vehicle in favor of a vehicle-sharing program, we assume that each driver averaged 4,019 miles per year, the average miles driven per vehicle-sharing member before joining a vehicle-sharing program (Martin and Shaheen, 2011). This provides us with a very conservative estimate because, as vehicle sharing grows more popular, it should spread to people who drive more miles than current joiners do, although perhaps not as many as the current average of 12,888 miles per driver (Santos et al., 2011, Table 23). (As noted in the upcoming section "Attractions of Vehicle Sharing Versus Private Ownership," other research has found various mileage "cut-off" points above which vehicle sharing would be more expensive than ownership, but these estimates all exceed 4,019 miles per year.)

We assume that each driver reduces VMT by 27 percent, or 1,081 miles.[4] In 2009, the average fuel economy of the U.S. light-duty vehicle fleet (cars and light-duty trucks) was 20.8 miles per gallon (EIA, 2011). We estimate the reduction in gallons of gasoline consumed made possible by substituting vehicle sharing for use of personally owned vehicles by dividing the estimate of fewer miles driven by the average fuel economy of the U.S. fleet. We then multiplied the gallons saved by 8.79, the number of kilograms of carbon dioxide released into the atmosphere when a gallon of gasoline is burned (Davis, Diegel, and Boundy, 2011, Table B-4, p. B-5), to calculate reductions in GHG emissions. We calculate that, for every driver who shifts usage from a personally owned vehicle to a vehicle-sharing program, 457 fewer kilograms of carbon dioxide would be emitted each year. Table 4.1 lays out these calculations.

This estimate is likely to be conservative because it assumes that the profile of a vehicle-sharing member will not change over time. Current models of vehicle sharing have been most effective in environments with medium- to high-density development, existing transporta-

---

[3] The study found a 27 percent reduction in actual VMT before and after adopting vehicle sharing and a 43 percent reduction if one considers forgone mileage (that is, miles that would have been traveled in a personally owned vehicle had the driver not joined a vehicle-sharing organization). To be conservative, we have used the observed reduction of 27 percent.

[4] Although Martin and Shaheen (2011) calculate emissions reductions per household, we use a simplifying assumption that each household has one vehicle-sharing member. Their study does not provide figures on the number of vehicle-sharing members in each household.

Table 4.1
Reductions in Annual Greenhouse Gas Emissions per Driver from Switching from Personally Owned Vehicles to Vehicle Sharing Due to Less Driving

Item	Source	Calculation	Value
Average number of miles driven annually by a North American vehicle-sharing member before joining a vehicle-sharing program (2008)	Martin and Shaheen (2011)		4,019
Reduction in miles driven from participating in a vehicle-sharing program (%)	Martin and Shaheen (2011)		27
Reduction in miles driven from participating in vehicle-sharing program (miles)		27% of 4,019	1,085
Average fuel economy of the U.S. light-duty fleet (2009) (mpg)	EIA (2011)		20.8
Reduction in gasoline consumption from participating in a vehicle-sharing program (gallons)		1,085 miles/ 20.8 mpg	52
GHG emissions from burning 1 gallon of gasoline (kg of CO_2e)	Davis, Diegel, and Boundy (2011)		8.79
Net savings in annual GHG emissions from reductions in driving (kg of CO_2e)		52 × 8.79	457

tion options (such as transit, walking, and bicycling), and particular demographics (members tend to be more educated, affluent, and technology savvy than the general population). People living in these neighborhoods are likely to drive less than the U.S. national average, as demonstrated by the difference between the VMT of a current vehicle-sharing member (4,019 miles) and an average driver (12,888 miles). If vehicle sharing becomes mainstream, reductions in GHG emissions from greater adoption of vehicle sharing would be greater than estimated above because those new members are likely to have driving habits more similar to the average.

Several European studies have found that trips on public transportation substitute for some of the decline in VMT by personally owned motor vehicles (Peter Muheim and Partner, 1998; Rydén and Morin, 2005).[5] The rest of the reduction in VMT is due to fewer trips or alternative means of transportation, such as walking, bicycling, or ride sharing. The energy consumed to carry these individuals on public transportation should be deducted from total energy savings. The amount of the reduction in energy use will depend on transit mode: On average, heavy rail consumes less energy per passenger-mile than bus. However, the information we have available for energy consumed per passenger-mile in the United States is based on the average, not the marginal rider.[6] Because transit service generally already exists in those neighborhoods most conducive to vehicle sharing, the additional energy consumption used to move an additional passenger by bus or rail is substantially less than the average amount

---

[5]   Peter Muheim and Partner, in a survey of Swiss vehicle-sharing members, find that, on average, vehicle kilometers traveled (VKT) fell by 9,400 (5,840 VMT) annually and that kilometers traveled on transit increased by 5,900 (3,665), meaning that about 65 percent of kilometers shifted to transit. (Another 17 percent shifted to walking, bicycling, or motorbike, and 20 percent of VKT were forgone—members' total VKT decreased) (Peter Muheim and Partner, 1998). Two surveys (in Bremen, Germany, and in Belgium) finds that members' VKT decreased by about 3,000 (1,864 VMT) and that transit use increased by about 1,100 VKT (683 VMT), meaning that about 36 percent of VKT shifted to transit (the amount that shifted to other modes, or that was forgone, was not discussed) (Rydén and Morin, 2005).

[6]   In the United States, a person traveling by bus uses an average of 4,315 Btu per mile, while a person traveling by rail uses an average of 2,577 Btu per mile (Davis, Diegel, and Boundy, 2009).

per passenger. Accordingly, we did not deduct additional energy use by mass transit from our totals because we lacked information about the additional energy needed for one additional passenger.

More-Efficient Vehicles. When an individual purchases his or her own vehicle, the buyer often selects a vehicle that will serve all the likely demands that the users may have for it. For example, if a family needs a minivan to go camping on weekends, the family may purchase a minivan, even though it rarely needs a vehicle of that size for day-to-day trips.

Vehicle sharing has the advantage of permitting drivers to choose the vehicle that is most appropriate for the intended use. With vehicle sharing, a family could choose a minivan for weekend camping trips, a truck to haul mulch, and a more fuel-efficient vehicle for shopping trips. The better fit between need and vehicle reduces fuel consumption and GHG emissions.

We do not have data on the specific shared vehicles driven to determine the extent of this shift on a trip-by-trip basis. However, a survey of vehicle-sharing members finds that vehicles acquired by those households (generally to replace existing vehicles) had a median fuel economy of 24 mpg, while the shared vehicles they generally drove had a median fuel economy of 31 mpg (Martin and Shaheen, 2010). This is consistent with results from other countries; a Swiss study finds that, in 2005, emissions of carbon dioxide per vehicle from the shared vehicle fleet from that country's main service provider were an average of 18 percent lower than those for all new cars sold in Switzerland and 25 percent lower than the total fleet (Haefeli et al., 2006).

We have incorporated these savings in fuel consumption and GHG emissions into our estimates of potential reductions in fuel use and GHG emissions from substituting vehicle sharing for use of personally owned vehicles. To capture these savings in our estimate, we assume that, prior to joining the program, participants achieved the average fuel economy of the existing light vehicle stock—20.8 mpg in 2009 (EIA, 2011). After joining the program, participants are assumed to choose smaller, more fuel-efficient vehicles for many trips than their previous personally owned vehicle. Vehicles in vehicle-sharing programs are generally used more intensively than personally owned vehicles because they are driven more times per day. Therefore, we expect that the average age of vehicles in fleets operated by vehicle-sharing programs is lower than in the U.S. fleet as a whole. Consequently, participants will be driving newer vehicles, on average, than when they drove personally owned vehicles. To capture this difference in usage and average age of vehicles, we assume that vehicles in vehicle-sharing fleets achieve the average fuel economy of new cars in the United States, 26.7 mpg in 2009, rather than the average for the entire U.S. light-duty fleet (EIA, 2011). We use the new-car average, rather than the new light-duty fleet average, on the grounds that vehicle-sharing members will take a greater proportion of trips in cars than in trucks and that the cars they drive are likely to be newer than personally owned vehicles.

Using these assumptions, we calculate that, rather than consuming 347 gallons of gasoline per year to drive a personally owned vehicle, a participant in a vehicle-sharing program would consume 270.3 gallons of gasoline. Because burning 1 gallon of gasoline emits 8.79 kg of GHG emissions, the savings in gasoline from participating in a vehicle-sharing program would result in a reduction in GHG emissions of 674 kg per year. Table 4.2 shows these calculations.

Fewer Vehicles Manufactured. Manufacturing motor vehicles uses substantial amounts of energy and releases large amounts of associated GHGs. According to Bandivadekar et al. (2008), manufacturing a typical car releases 7.7 metric tons of GHG emissions and a light

Table 4.2
Reductions in Annual Greenhouse Gas Emissions per Driver from Using More-Efficient Vehicles Due to Participating in a Vehicle-Sharing Program

Item	Source	Calculation	Value
Average number of miles driven annually by a North American vehicle-sharing member before joining a vehicle-sharing program (2008)	Martin and Shaheen (2011)		4,019
Reduction in miles driven from participating in a vehicle-sharing program (%)	Martin and Shaheen (2011)		27
Total number of miles driven by a participant in a vehicle-sharing program			2,934
Average fuel economy of the U.S. light-duty fleet (2009) (mpg)	EIA (2011)		20.8
Gasoline consumption assuming average fuel economy (gallons)		2,934 miles/ 20.8 mpg	141.1
Average fuel economy of a new U.S. car (2009)[a] (mpg)	EIA (2011)		26.7
Gasoline consumption from participating in a vehicle-sharing program (gallons)		2,934 miles/ 26.7 mpg	109.9
Reduction in gasoline consumption between a personally owned vehicle and a vehicle in a vehicle-sharing program (gallons annually)		141.1 − 109.9	31.2
GHG emissions from burning 1 gallon of gasoline (kg of CO_2e)	Davis, Diegel, and Boundy (2011)		8.79
Net savings in annual GHG emissions from reductions in driving (kg of CO_2e)		31.2 × 8.79	274

[a] We assume that, on average, vehicles in vehicle-sharing programs achieve average fuel efficiency for new cars in the United States.

truck releases 10 metric tons, roughly one-third of the life-cycle GHG emissions from light-duty vehicles.

Vehicle sharing should lead to fewer vehicles being manufactured. First, shared vehicles are used much more intensively than personally owned motor vehicles. Shared vehicles, such as taxicabs, are driven more miles over their service lives than personally owned vehicles. Consequently, fewer vehicles are needed for a given number of VMT because the existing fleet is used more intensively. Second, because members drive fewer miles in aggregate, fewer vehicles are needed to provide the reduced number of VMT. Both factors result in fewer vehicles in the fleet and, therefore, fewer vehicles manufactured. Because vehicle manufacturing consumes energy, having fewer vehicles manufactured results in lower emissions of GHGs.

According to Lu (2006), the estimated average service life of U.S. light vehicles is 152,137 miles for cars and 179,954 miles for light trucks. Using these data, we estimate that the average service life of light-duty vehicles in the United States is 164,007 miles.

Taxicabs and other heavily used commercially operated motor vehicles have longer average service lives than personally owned vehicles. Although definitive figures were not available, a 2006 report on taxis in New York City finds that the average taxicab traveled 64,600 miles per year and had to be taken out of service after five years, implying a service life of at least 323,000 miles before being retired (Schaller Consulting, 2006).[7]

Using these figures, we find that 49 percent fewer vehicles would need to be manufactured to provide an equivalent number of vehicle miles in vehicle-sharing programs as personally owned vehicles because vehicles in vehicle-sharing programs are driven much farther over the course of their service lives. Because manufacturing vehicles uses energy and emits GHGs, fewer vehicles purchased because of greater use would result in a commensurate reduction in energy use and GHG emissions—in this case, a reduction of 49 percent, or 105 kg of CO_2e, per participant on an annualized basis. Our analysis is shown in Table 4.3.

Second, as noted earlier, members in vehicle-sharing programs drive 27 percent fewer VMT than individuals who own their own automobiles. Consequently, 27 percent fewer vehicles are needed to provide transportation for a group of participants because they drive less. As a consequence, vehicle-sharing services would reduce energy use and GHG emissions due to manufacturing motor vehicles by 27 percent compared with personally owned motor vehicles, or by 57 kg of CO_2e annually.

Using this figure and the arguments outlined earlier, we show our calculations of net reductions in GHG emissions in Table 4.3. Reductions are equivalent to a total of 162 kg of CO_2e annually.

Table 4.4 sums the reductions in emissions from each of these factors leading to lower fuel consumption and GHG emissions from vehicle-sharing services. Reductions per participant are large, on the order of 893 kg of CO_2e annually. The average light-duty vehicle driven by a potential vehicle-sharing member emits 1,698 kg of CO_2e annually from driving.[8] An additional 8,681 kg is emitted during the manufacture of the vehicle. If the emissions from manufacturing are spread out over the average 12.7-year lifetime of the vehicle (682 kg per year), total average annual emissions per vehicle are 2,380 kg. So, the average vehicle-sharing member would emit only 1,487 kg of CO_2e per year, a reduction of 37 percent.

These reductions are somewhat greater than the estimates developed by Martin and Shaheen (2010), 840 kg of CO_2e per year, and Haefeli et al. (2006), 290 kg of CO_2e per year. These differences are driven primarily by our assumption about the fuel economy of the vehicles replaced (U.S. light-duty vehicle fleet average) and the inclusion of savings in GHG emissions from use of more-efficient vehicles and fewer vehicles manufactured.

---

[7] From a different source, a Canadian newspaper notes that, for Lincoln Town Cars, often used by livery services, "a 400,000 mile service life is typical, some Town Cars reportedly reaching as much as 600,000 to 700,000 miles" (Moore, 2011).

[8] Calculation based on 4,019 annual VMT, divided by average fuel economy of 20.8 miles per gallon, multiplied by 8.79 GHG emissions in 1 gallon of gasoline.

Table 4.3
Potential Reductions per Driver in Annual Greenhouse Gas Emissions Stemming from Reduced Demand for Motor Vehicles

Item	Source	Calculation	Value
Analysis of reductions in GHG emissions from using vehicles more intensively			
Average service life of personally owned motor vehicle (miles)[a]	Calculated from Lu (2006)		164,007
Average service life of commercially operated vehicle (taxicab) (miles)	Schaller Consulting (2006)		323,000
Average number of miles driven annually by North American vehicle-sharing members (2008)	Martin and Shaheen (2011)		4,019
Number of personally owned vehicles utilized annually to drive 4,019 miles		4,019 miles/ 164,007 miles	0.0245
Number of vehicles in vehicle-sharing programs utilized annually to drive 4,019 miles		4,019 miles/ 323,000 miles	0.0124
Annual net reductions in the number of vehicles per driver			0.0121
GHG emissions per vehicle manufactured (kg of CO_2e)[b]	Calculated from Bandivadekar et al. (2008) and EIA (2011)		8,681
Subtotal: Reduction in GHG emissions from requiring fewer vehicles per driver because of greater use of vehicles (kg of CO_2e)		0.0121 vehicles × 8,681 kg of CO_2e	105
Reductions per participant in vehicle-sharing programs in annual GHG emissions stemming from reduced demand for motor vehicles because of fewer miles driven			
Number of personally owned vehicles utilized annually to drive 4,019 miles		4,019 miles/ 164,007 miles	0.0245
Reduction in miles driven from participating in a vehicle-sharing program (%)	Martin and Shaheen (2011)		27
Reduction in number of vehicles required annually in vehicle-sharing programs due to reduced driving		0.27 × 0.0245 vehicles	0.0066
GHG emissions per vehicle manufactured (kg of CO_2e)	Calculated from Bandivadekar et al. (2008) and EIA (2011)		8,681
Subtotal: Reduction in GHG emissions from requiring fewer vehicles per driver due to less driving (kg of CO_2e)		0.0066 vehicles × 8,681 kg of CO_2e	57
Total reductions in GHG emissions from requiring fewer vehicles per driver (kg of CO_2e)		105 + 57	162

[a] The average service life of U.S. light vehicles has been estimated at 152,137 miles for cars and 179,954 miles for light trucks (Lu, 2006). The mix of cars and light trucks in the U.S. fleet in 2009 was 131.33 million cars and 97.76 million light-duty vehicles (EIA, 2011). Using these figures, we estimate the average life of a light-duty vehicle in the United States at 164,007 miles.

[b] As noted earlier, Bandivadekar et al. (2008) estimates that manufacturing a typical car releases 7.7 metric tons of GHG emissions and that a light truck releases 10 metric tons. Weighting these amounts by the share of cars and light trucks in the U.S. light-duty fleet cited in the previous note, average annual emissions per vehicle are 8,681 kg.

Table 4.4
Potential Annual Reductions in Greenhouse Gas Emissions per Participant from Participating in Vehicle-Sharing Services

Source of Reduction	Source	Kg of GHG Emissions Avoided
Reduced VMT	Table 4.1	457
Use of more-efficient vehicles	Table 4.2	274
Reductions in the number of vehicles manufactured due to use of vehicles that is more intensive and reduced VMT	Table 4.3	162
Total		893

Demand for Vehicle Sharing

This section discusses the existing markets for vehicle sharing and the incentives and barriers for wider use.

Vehicle-Sharing Markets

Vehicle sharing serves different markets:

- *Neighborhood-based vehicles.* The earliest and still most prevalent model for vehicle sharing is based on locating vehicles in moderate- to high-density neighborhoods well-served by public transit.[9] The vehicles provide a convenient way to make trips that cannot easily be made by other modes; density makes it possible to locate vehicles within walking distance of members.

- *Transit-station vehicles.* Many vehicle-sharing organizations have placed vehicles at transit stations. Locating vehicles at stations permits some members to make trips by transit when the destination is too far from the station to reach by walking. In transportation planning, this is known as the *last-mile problem*—transit can get a rider to an area but not to the destination itself.

- *Low-income groups.* Some vehicle-sharing organizations have worked with social service providers to make vehicle sharing attractive and affordable to people with low incomes. City CarShare in San Francisco extends reduced-cost membership to residents of buildings managed by affordable housing developers, and to CalWorks (the state welfare agency) recipients (City CarShare, 2009).

- *Businesses and governments.* Firms and government agencies may join vehicle-sharing organizations so as to provide their employees with access to vehicles during work hours for business purposes. Flexcar (now merged with Zipcar) established a business membership program in 2002; Zipcar followed with its own program in 2004. These programs can replace company cars or serve as an incentive to employees to not drive to work. Some city governments have replaced their own fleet vehicles with vehicles managed by vehicle-sharing organizations (Shaheen, Cohen, and Chung, 2009).

---

[9] Although there are no hard-and-fast definitions of moderate- and high-density neighborhoods, in the literature, *moderate density* generally means anything from eight to 20 housing units per acre (small single-family houses, row houses, or garden apartments); *high density* is upward of 20 housing units (row houses or apartments of four stories or more).

- *College and university campuses.* Campuses have been an important market for vehicle-sharing services: As of November 2010, Zipcar had vehicles at more than 225 college campuses (Zipcar, 2011). Many universities have severely restricted space for parking. Moreover, student populations tend to be a receptive audience for environmentally friendly ideas. Students also tend to have lower incomes than nonstudents, so the trade-off between owning a vehicle or renting may look more attractive than for individuals with greater means. Some universities have experimented with vehicle sharing for a considerable period of time (Bonner, 2010).

Vehicle sharing has been a harder sell outside these well-defined markets. Commuters who drive to work would not find vehicle sharing a realistic option because the per-trip costs of shared vehicles would be prohibitively expensive. Given that, in 2000, there were about 128 million workers in the United States, of whom 97 million drove to work alone (another 15 million drove in carpools) (Pisarski, 2006), this suggests a large swath of the U.S. population for whom owning a vehicle is probably the most viable option.

We did not find any successful models of suburban or rural vehicle-sharing programs. The low-income programs mentioned earlier have been in urban areas (Ortega, 2005). A group that works in the Central Valley in California wrote a report about prospects for vehicle sharing for residents of small towns and agricultural areas but has not pursued development of such options (Higginbotham, 2000; Moffat, 2010). The only viable models in small towns outside of major metropolitan areas at this time seem to be at college or university campuses and communities expressly formed for environmental reasons (Dancing Rabbit Ecovillage, undated).

Attractions of Vehicle Sharing Versus Private Ownership

Demand for vehicle-sharing services is driven by a variety of factors. In this section, we investigate key drivers.

Cost. Vehicle-sharing organizations tend to charge by the hour, not the mile. For vehicle-sharing organizations throughout the United States, the cost for a typical four-hour trip ranges from $20 to $48 (Carsharing.net, undated). The American Automobile Association (AAA) estimated that the average cost to drive 10,000 miles in a personally owned vehicle in 2009 was $7,393 ($0.739 per mile), although this varies with the size and fuel efficiency of the vehicle, the insurance coverage, the fees and taxes levied in a particular jurisdiction, the age of the vehicle, and whether it is owned outright or financed (AAA, undated). The per-mile cost falls with the distance driven.

Millard-Ball et al. (2005) find that vehicle sharing was cheaper than ownership for drivers who drove 5,000 vehicle miles or less per year. Higginbotham finds that drivers who drove 8,000 miles per year or less also came out ahead (Higginbotham, 2000). Litman finds a fairly wide range of 4,000 to 10,000 miles driven per year below which vehicle sharing is cheaper than owning one's own vehicle (Litman, 1999, Figure 2). Three surveys of U.S. vehicle-sharing members find that members reported saving between $154 and $435 monthly in transportation costs (Shaheen, Cohen, and Chung, 2009).

Hassles of Vehicle Ownership. Vehicle-sharing members often cite the less pleasant aspects of vehicle ownership, such as maintenance, as a reason for choosing vehicle sharing. Parking is another incentive, especially for households that do not have access to convenient or inexpensive parking.

Demographic and Attitudinal Changes. Vehicle sharing combines an environmental ethos with the ability to match vehicles to specific purposes, attributes that seem to appeal to the younger, urban, and well-educated membership of vehicle-sharing organizations (see Douma and Andrew, 2006, for a discussion of the demographics of vehicle-sharing members).

Barriers to Adoption

Despite rapid growth in the past decade in the United States, membership in vehicle-sharing programs is still tiny in percentage terms: only 0.2 percent of the 200 million licensed drivers in the United States. Despite Europe's longer experience with car sharing and tighter land-use patterns, adoption rates in Europe are similar to those in the United States. An estimate of vehicle sharing in Germany found that 0.16 percent of licensed drivers belonged to shared-vehicle organizations (Loose, Mohr, and Nobis, 2006). Growth in membership has been constrained by a variety of factors.

Lack of a Proven Business Model. Vehicle-sharing organizations have a mixed track record as business concerns. Zipcar went public in April 2011; however, despite substantial increases in membership, it has yet to turn a profit (Bieszczat and Schwieterman, 2011). The difficulty in creating a profitable business may explain why many other vehicle-sharing organizations are cooperatives or not-for-profits; of more than two dozen vehicle-sharing organizations in the United States, fewer than half are for-profit (Carsharing.net, undated). The largest European organization, Mobility CarSharing in Switzerland, operates as a cooperative; in 2009, it earned a profit of 1.4 million Swiss francs (about $1.2 million) (Mobility CarSharing, 2010).

In the United States, the market is dominated by Zipcar. Most other vehicle-sharing services are small, and many have closed; by one estimate, 17 vehicle-sharing organizations have opened and then closed since 1994. Some of these were pilot projects with a sunset date; about half were either merged with other programs or closed because of "organizational deficits and greater staffing needs" (Shaheen, Cohen, and Chung, 2009). It is not clear whether growth may come from new organizations or existing small organizations or whether Zipcar will continue to dominate.

Lack of Awareness or Interest. Surveys have found that knowledge about vehicle sharing is low and that, even when it is explained, some people do not like the idea. A German survey finds that only 15 percent of people recognized the term *car sharing*, although vehicle sharing had operated in the country for more than ten years. Almost three-quarters of respondents agreed with the statement, "I prefer using my car to sharing a car with other people"; fewer than half "can imagine sharing a car with other people" (Loose, Mohr, and Nobis, 2006). These barriers, particularly the attitudinal, have been difficult to overcome.

Receptivity. Several studies note the difference between the potential for vehicle sharing based on objective criteria, such as those who could commute by transit, and subjective criteria, such as receptivity to vehicle sharing. For example, a Swedish study estimates a "theoretical potential" of 25 percent of the population as potential members of vehicle-sharing programs, while only 5.6 percent were really potential customers according to market research of consumer interest (Vägverket, 2003).

Availability of Government Support. Vehicle sharing can be and has been helped by various levels of government entities: local governments, transit agencies, and metropolitan planning organizations. Support can include provision of parking spaces, changes in parking regulations, start-up subsidies, and outreach through commuter assistance programs. Although none of these is strictly necessary to foster vehicle sharing, lack of such support may hinder efforts to make vehicle sharing more widespread.

Lack of Convenient Parking. The current model of vehicle sharing requires dedicated parking spaces. Limited parking can serve as a barrier to the expansion of vehicle sharing. Although off-street parking would seem to be a natural choice, Shaheen, Cohen, and Martin (2010) find that prospective members are more likely to join when the vehicles are visible.

High Taxes. In a 2011 study of 91 areas with vehicle sharing, Bieszczat and Schwieterman find that taxes levied on shared vehicles are, on average, close to 18 percent for a one-hour rental, which far exceeds the average sales tax rate of 8 percent in those cities. The tax rate on conventional car rental (for a 24-hour period) is almost 16 percent. (Both figures are weighted by population.) In eight of the 25 largest cities, tax rates for a one-hour rental exceed 30 percent. Many cities charge flat rates on every rental transaction, regardless of length, in addition to other vehicle rental taxes.

Hassles of Vehicle Sharing. For vehicle sharing to be attractive, members must follow rules, such as not smoking or leaving trash in vehicles, and the technologies that make vehicle sharing possible must work reliably. Examples of the problems that can occur with vehicle sharing are highlighted in an article and video prepared by reporters for the *Austin American-Statesman* (Wear, 2010).

Difficulties in Obtaining Insurance. When vehicle sharing was introduced, providing automobile insurance to cover members was a major barrier. In the early 2000s, insurance accounted for 20 to 48 percent of operating costs. By 2005, half of vehicle-sharing organizations reported that obtaining insurance was a problem. In part, this was due to the expansion of the service to student populations, who are generally under 25 and are riskier drivers, on average, than older drivers.[10] Because they are riskier, students are charged much higher premiums. Some expect insurance costs to decrease as insurance companies become more familiar with vehicle sharing and adapt pricing models based on more data. However, insurance costs may continue to be a barrier in some markets. Pay-as-you-drive insurance models, which are slowly being introduced in the U.S. market, may help in this respect (Shaheen, Cohen, and Chung, 2009).

Land-Use Patterns. In the United States, many people live in neighborhoods where population densities may be too low to support conventional vehicle sharing. In suburban and rural areas, with multicar garages and ample free parking at most destinations, almost all households already own vehicles. One study finds that more than 97 percent of rural households own at least one vehicle, compared with 92 percent of urban households (Pucher and Renne, 2005). It is difficult to square dispersed land use and high vehicle ownership with a model that assumes that shared vehicles will be within walking distance from a residence or workplace.

---

[10] A 2008 survey finds that vehicle-sharing firms in the United States that provided vehicles on university campuses paid an average premium of almost $2,500 per vehicle, while those that did not serve campuses had an average premium of about $1,500. In contrast, Canadian vehicle-sharing services pay as little as 600 Canadian dollars (about 585 U.S. dollars) per vehicle for insurance. The lower Canadian rates reflect the greater incidence of public insurers in Canada (Shaheen, Cohen, and Chung, 2009).

Potential Demand for Vehicle Sharing

In this section, as in Chapter Three, we ask how much GHG emissions could be reduced if U.S. drivers were to participate more extensively in vehicle-sharing services. We compare actual participation in vehicle-sharing programs with three cases that vary in terms of market penetration, defined as the percentage of all drivers who are vehicle-sharing members. The first case is the status quo, or base case. In the second, a higher percentage of individuals with characteristics similar to those of current users join vehicle-sharing services. In the third case, we assume much greater participation in vehicle-sharing services as a consequence of participation by groups that currently do not participate in these programs. The assumptions about market penetration are based on estimates from the literature, also discussed in this section.

Table 4.5 compares three sets of estimates of potential demand for vehicle sharing in the United States. Zhao's estimate appears to be based on commercial market research by the firm Frost and Sullivan, which we were unable to review because the study is proprietary.[11] Schuster et al. (2005) does not look at the demographic base for potential members but compares the costs of owning and sharing vehicles. They assume that, if vehicle sharing is cheaper than owning a vehicle, the individual vehicle owner is a potential candidate for the service. If not, the individual is assumed to prefer a personally owned vehicle. The authors acknowledge that attitudes, not just financial concerns, determine the decision to own rather than share. Shaheen, Cohen, and Roberts (2006) used surveys of vehicle-sharing operators to arrive at a figure for potential members of 12.5 percent of all residents over 21 years of age in major metropolitan areas (those that currently have vehicle-sharing programs or are likely to in the future).

Another study looks at the characteristics of neighborhoods where vehicle sharing has been viable. The authors suggest that four characteristics can help predict which neighbor-

Table 4.5
Estimates of Potential Vehicle-Sharing Membership in the United States

Source	End Year of Analysis	Millions of U.S. Members	Basis for Estimate
Zhao (2010)	2016	4.4	Methodology not available
Schuster et al. (2005)	N/A; market potential	6–7.5[a]	Owners of 3.6% to 4.5% of vehicles in an MSA would find it cheaper to share than to own
Shaheen, Cohen, and Roberts (2006)	N/A; market potential	20.3[b]	12.5% of everyone over 21 years of age in major metropolitan areas, based on surveys

NOTE: N/A = not applicable. MSA = metropolitan statistical area.

[a] We estimated the figure of 6 million to 7.5 million based on Shuster et al. (2005). The 2009 NHTS finds that 171.8 million vehicles were located in MSAs. If the range of 3.6 to 4.5 percent is used across MSAs, then as many as 6.2 million to 7.7 million vehicles could be replaced by shared vehicles. Assuming that each forgone vehicle was used by 0.97 persons (the ratio of drivers to light-duty vehicles is 0.97, according to the NHTS), the potential for vehicle sharing ranges from 6 million to 7.5 million people.

[b] To derive the figure of 20.3 million, we calculate based on census estimates for 2009 that 229 million people currently live in major metropolitan areas and that 71 percent of the U.S. population is over the age of 21 (U.S. Census Bureau, 2009c).

---

[11] Although it is available for purchase, this report is prohibitively expensive.

hoods will be most amenable to vehicle sharing: the percentage of one-person households, commute mode shares (lower-than-average drive-alone shares and higher-than-average walk shares), vehicle ownership (the percentage of households with zero vehicles or one vehicle) and housing units per acre. The analysis characterized the eventual viability of vehicle sharing as either low or high but did not suggest what percentage of households in such neighborhoods might join vehicle-sharing organizations (Celsor and Millard-Ball, 2007). Despite the usefulness of this analysis for identifying neighborhoods where vehicle sharing may be viable, the methodology does not lend itself to estimating the future numbers of vehicle-sharing members.

We provide a cautionary note that estimates of growth in vehicle sharing have previously been proven wrong. A 1994 study predicted that the market potential in Germany was 2.45 million members; however, ten years later, the market stood at 70,000 (Loose, Mohr, and Nobis, 2006). More than a decade ago, observers noted that actual membership rates were only 3 to 8 percent of projections of membership levels from a decade earlier (Harms and Truffer, 1998).

Potential Cases for Greater Market Penetration of Vehicle Sharing

To investigate the potential contribution of vehicle sharing for reducing GHG emissions, we consider three cases based on current data:

- Base case: This case assumes the current number of vehicle-sharing members, 560,000.
- Supportive policy case: This case assumes the higher end of the Schuster et al. (2005) estimates, with 4.5 percent of drivers adopting vehicle sharing.
- Maximum adoption case: This case is based on the Shaheen, Cohen, and Roberts (2006) estimate that 12.5 percent of drivers over 21 in metropolitan areas join vehicle-sharing organizations.

Using these cases and the analysis of net reductions in GHG emissions from shifting from use of personally owned vehicles to vehicle sharing, we calculate potential reductions in GHG emissions under the supportive policy and maximum adoption cases (Table 4.6). The base-case reductions in GHG emissions are those that have already been realized because of the growth in vehicle sharing. As can be seen, under these assumptions, vehicle sharing might reduce transportation-related GHG emissions by 0.05 to about 1.7 percent, depending on the types of policy, behavior, and business-model changes that occur in the future. For each case, we compare the reduction in emissions in metric tons with the base case. Instead of predicting future growth rates, we use assumptions based on 2009 conditions.

In addition to reducing GHG emissions, vehicle sharing is also likely to provide other benefits, such as reducing congestion because of fewer vehicles both on the road and parked on city streets. It should also reduce the need for parking and reduce congestion-related increases in travel time and emissions of conventional air pollutants. Because many transportation problems can be improved by small changes in the number of vehicles using the roads, these benefits may be particularly valuable in large metropolitan areas.

Table 4.6
Potential Reductions in Greenhouse Gas Emissions from Switching from Personally Owned Vehicles to Vehicle Sharing Under Three Cases

Characteristic	Base Case	Supportive Policies Case	Maximum Adoption Case
Number of participants	560,000	7,500,000	20,300,000
Assumed percentage adoption by drivers in urban areas	0.27	4.5	12.5
Reductions in GHG emissions per participant (kg of CO_2e)	893	893	893
Total reductions in GHG emissions (millions of metric tons)	0.5	6.7	18.1
Total GHG emissions from light-duty vehicles (2009) (millions of metric tons)	1,071	1,071	1,071
Percentage reduction from GHG emissions from U.S. light-duty fleet	0.05	0.6	1.7

NOTE: Total emissions from the light-duty fleet were calculated from data on total GHG emissions from light-duty vehicles (EIA, 2012, Table 19).

Policy Options

In our view, three criteria are necessary to greatly expand vehicle sharing:

- The service has to be either cheaper or more convenient than owning a personal automobile, preferably both.
- Providers have to make a profit or at a minimum cover costs.
- Penetration rates have to reach a critical mass in a geographic area, so that operators find it attractive to operate there.

In this section, we consider how the cost and convenience of vehicle sharing can be influenced by new policies, as well as by changes in technology, markets, and behavior. The information in this section is based on existing uses, pilot programs, and ongoing research. The main factor that unites them is new technologies, especially technologies that provide real-time information through the global positioning system and allow users access from a variety of platforms. These assumptions ground these suggestions in reality but, at the same time, suggest that other factors besides the availability of technology may be at work. Even if the technology exists, it may not reduce costs or increase convenience, key reasons Americans rely on personally owned vehicles.

Technological changes are the most conventional area for DOE to address, and DOE might carry out research on some of the technologies that could speed adoption. However, technology is not the only aspect of vehicle sharing to which DOE and other parts of the federal government might devote further study. The market for many of these services is unproven and may well be subject to the same forces that have kept current forms of vehicle sharing from expanding outside of niche markets. It is extremely difficult for any service to overcome the convenience of personal vehicle ownership (see Poudenx, 2008).

Adopt Parking Policies That Allow One-Way Dynamic Vehicle Sharing

Almost all currently operating vehicle-sharing services require that trips be reserved in advance and that vehicles be returned to the same parking spaces where they were picked up. Under a one-way dynamic model, the driver locates a vehicle in real time and drives it to his or her destination, paying only for a one-way trip. The vehicle can be parked in any available unpaid space. The location of the vehicle is continuously updated via wireless communications. This model can reduce costs because drivers do not need to pay for time they are not using the vehicle. It can increase convenience because it allows one-way trips, which are currently not allowed in conventional vehicle sharing, and because drivers do not have to specify a return time in advance.

Parking policy changes are important because the service cannot operate effectively without a large number of parking spaces available throughout an area. When spaces are restricted to neighborhood residents through a permit program, or when they are paid in short time increments at meters, they would not be available to park a shared vehicle indefinitely.

At the time of this writing, car2go operates programs along this model in Ulm and Hamburg, Germany; Vienna, Austria; Amsterdam, Holland; Lyon, France; Austin, Texas; San Diego, California; and Vancouver, Canada. This model is quite new (the Ulm program, the first, dates to October 2008; in Austin, the program was piloted in November 2009 and launched throughout Austin in May 2010), so it is not clear whether one-way dynamic vehicle sharing constitutes a viable model. One study suggests that car2go might reduce members' emissions on average by 146 to 312 kg per year (Firnkorn and Müller, 2011). Car2go is a subsidiary of Daimler AG, the car manufacturer, so presumably the parent company can permit the operation to run at a loss for a period while it investigates workable business models.

One-way vehicle sharing would likely appeal to the same markets as conventional vehicle sharing; however, by increasing convenience and decreasing costs, it may speed adoption within urban areas. Technologies and policies that might help make this service more attractive could include better computer applications for finding vehicles, algorithms that are more sophisticated for predicting where cars are needed, cheaper or less polluting ways to redistribute vehicles if there are too many in some areas and not enough in others, and more unrestricted parking.

Change Insurance Regulations to Facilitate Peer-to-Peer Vehicle Sharing

Peer-to-peer vehicle sharing allows members to share vehicles among themselves. Members rent their personally owned vehicles to each other in exchange for a commission. Members who own vehicles maintain them and set an hourly rate. A vehicle-sharing organization would still be needed in this model: It would install devices in the vehicle so that owners and renters do not need to exchange keys. It would also maintain the rental system that confirms reservations (Hamilton, 2010).

One obstacle to this form of vehicle sharing is insurance regulations that forbid policyholders from renting out their vehicles (Cabanatuan, 2010). Although this problem has been resolved in those locations that have peer-to-peer vehicle sharing, insurance regulations vary by state. State insurance regulations could be changed so as to allow coverage for private rental.

As of this writing, both California (Assembly Bill 1871) and Oregon (House Bill 3149) have adopted legislation that allows peer-to-peer vehicle sharing (Roth, 2010; Witkin, 2011a).

It is too soon to speculate about how successful this form of vehicle sharing might become. Some advantages are that the number of potential users may be larger than that of conventional vehicle sharing because the pool of available vehicles can be much larger and more dispersed and the costs of peer-to-peer models are likely to be lower than renting vehicles through conventional vehicle sharing. Also, the ability to earn money from one's personal vehicle may draw people in, as might the number of people who occasionally telework (i.e., work off-site using telecommunication technology to communicate) and therefore do not need a car to commute every day. On the downside, it is not clear how many people would want to participate in a model in which strangers drive their cars.

Peer-to-peer vehicle sharing has been launched in a few cities. It was first implemented by DriveMyCar Rentals in Australia in December 2008 (DriveMyCar Rentals, undated). Other organizations include WhipCar in the United Kingdom, which opened in London in April 2010 with plans to launch throughout the United Kingdom (Insley, 2010), and RelayRides, which launched in Boston and Cambridge, Massachusetts, in June 2010 (Belson, 2010) and has since expanded to San Francisco (Abdel-Razzaq, 2011). Other firms that have or plan to launch similar models are Getaround, Wheelz, JustShareIt, and Spride.

Help Advance Technologies That Facilitate Vehicle Sharing

One appealing characteristic of vehicle sharing is the ability to locate vehicles on short notice. Improving the technologies that allow such last-minute usage can make them available to a wider number of people. Currently, members of some vehicle-sharing firms can locate vehicles through a website or an iPhone but not using other smartphones (Car2Go, undated).

Another technology to advance peer-to-peer vehicle sharing is linking it with social networking sites to allow only known persons to drive a member's vehicle. Vehicle owners may be reluctant to loan out their vehicles out of concern they will be returned late, dirty, damaged, or not at all. Some attitudes may change if and when owners learn of positive experiences (in a manner similar to users of online retailer eBay providing positive feedback on buyers and sellers) and if owners are allowed to prohibit renting to certain members.

Alleviate the Tax Burden on Shared Vehicles

As noted in Bieszczat and Schwieterman (2011), in many cities, shared vehicles are subject to the same taxes and fees as conventional rental vehicles, leading to per-hour tax rates that far exceed sales tax rates. This tax burden could be alleviated through legislation to exempt shared vehicles or through tax credits.

Legalize and Provide Better Operating Environments for Shared Neighborhood Vehicles

Shared neighborhood vehicles are small, highly efficient vehicles used for short errands. Neighborhood vehicles could be motorized scooters, Segways, or other one- or two-person vehicles. Because they are designed to be used exclusively within a neighborhood, they do not need to have a long range or reach high speeds.

The service these vehicles could best provide is leisure. One analysis predicted that neighborhood electric vehicles might be most commonly used in such areas as golf courses and leisure communities that cater to older adults, island and resort communities with limited vehicular traffic and perhaps a high number of second homes, and transportation within institutions, such as large campuses or national parks (Morrison, 2002). One characteristic these environments share is a high concentration of leisure trips on lower-speed roads.

As of 2008, neighborhood vehicles were legal in 40 states, so legislation would be needed to allow them in all states. For safety reasons, most states restrict their operation to roads with speed limits of 35 mph or less (Fazzalaro, 2008). Other policies that could help speed adoption of neighborhood vehicles include dedicated parking and street patterns that allow vehicles to travel widely without the use of high-speed streets.

Such vehicles are already in use in some areas. The South Bay Cities Council of Governments in the Los Angeles region ran an 18-month pilot, the Local Use Vehicle (LUV) Project, in which individuals were loaned an LUV for three to six months and reported on their experiences. The 100 percent electric vehicles travel up to 25 miles per hour and are limited to use on streets with speed limits of 35 mph or below. They can travel up to 30 miles on one charge (California State Association of Counties, 2010).

Research Material Advances to Enable Greater User of Shared Neighborhood Vehicles

Some potential drawbacks of shared neighborhood vehicles include lower crashworthiness (because neighborhood vehicles do not need to have the same safety features conventional vehicles do) and shorter ranges than conventional vehicles (Fazzalaro, 2008). In addition, vehicles may not perform well in heavy rain or snow. Advances, such as strong, lightweight body materials, might help alleviate some of these concerns. The National Highway Transportation Safety Administration adopted safety standards specifically for neighborhood vehicles in 1998 (49 CFR § 571.500).

Develop Better Technology for Ride-Matching Services to Enable Smart Paratransit and Dynamic Ride Sharing

Although they are not vehicle sharing, smart paratransit and dynamic ride sharing provide transportation services that fulfill such needs as subsistence, participation, and leisure. Smart paratransit is a service in which riders request a ride from one location to another and are quickly matched with a vehicle that will pick them up at or near their location and drop them off at or near their destination. They would likely share that ride with one or more other passengers in a vehicle probably owned and operated by a commercial firm. Dynamic ride sharing is a similar concept but with the underlying assumption that drivers would provide rides in their own vehicles to other passengers.

The improvement of ride-matching technology would make these models far more efficient and, thus, attractive than they have previously been, and indeed some innovations have been introduced. The first fully autonomous paratransit dispatching system was started in Corpus Christi, Texas, in 2003. The dispatch system is linked to on-board vehicle computers that determine pickup and drop-off times. The only human input to the system is the requests made by customers, who can submit them online or by phone. However, although the technology functions satisfactorily, user needs were not satisfied, and community acceptance has not been as high as hoped (SAIC, 2005). Other studies have found that many operators of computer-assisted scheduling and dispatching (CASD) systems do not utilize optimization features, efficiency has not dramatically increased, and training issues have caused implementation problems. CASD trades off efficiency against ride time; the more passengers who are accommodated, the longer the trips take (Kessler, 2004). Improvements in these areas could lead to shorter wait times and travel times.

Unlike other vehicle-sharing services, smart paratransit and dynamic ride sharing might be used for work trips. Both models are efficient at serving many origins and destinations. Carpools and vanpools, both standing and "casual,"[12] already work in many areas, particularly in areas with incentives, such as high-occupancy vehicle (HOV) lanes or tolls. These innovations might overcome some obstacles that commuters face in forming carpool and vanpools, such as incompatible schedules, the need to drive when the usual driver is not available, and the difficulty of finding occasional rather than standing carpool partners. Some commercial firms have begun offering ride-share matching based on social networks, while others are using a dynamic ride-sharing model (Chan and Shaheen, 2011).

Advances in technology, although important, may not be sufficient to overcome other obstacles to dynamic ride sharing. In addition to legal issues about payment between individuals, other issues include concerns about privacy and safety, lack of flexibility, trip length and inconvenience, the difficulty of reaching a critical mass of users, and lack of incentives to participate (Hartwig and Buchmann, 2007). An analysis of a six-year attempt to build a dynamic ride-sharing service in the San Francisco Bay area identified several barriers: The service found it difficult to reach critical mass because there was no reward for trying the service if no match was found; parking spaces were plentiful, so the incentive to carpool was limited; and marketing was difficult without an established organization supporting the program (Kirshner, 2007).

Continue Research into and Regulation of the Technology to Enable Driverless Vehicles

Driverless vehicles can drive themselves. Driverless vehicles might be used in several ways in a vehicle-sharing framework: They could drive themselves to a rider's door on order, or they could serve as "personal rapid transit," originating at a station and driving to the rider's destination.

Most of the uses currently envisioned for driverless vehicles are in such environments as airports and college campuses (Dodson, 2007); however, in the more distant future, they could be used for transporting a wider range of drivers, especially groups with higher crash risks (e.g., teenagers, inebriated drivers).

Although certain intelligent driver-assistance technologies that make some decisions for drivers have already been commercialized—such as adaptive cruise control—driverless vehicles have been tested but not put into service.[13] Masdar City, the zero-carbon city under construction in the United Arab Emirates, is testing a personal rapid-transit system of underground, six-seat electric vehicles that operate with magnetic guidance (Arnott, 2010). In April 2011, London's Heathrow Airport launched a fleet of driverless electric vehicles that connect a parking lot to a terminal (Witkin, 2011b).

Both of these applications involve fixed guideways; the vehicles do not operate on ordinary streets, which would be necessary in a widespread system of driverless vehicles. Additional technological advances are needed to ensure the safe operation of driverless vehicles in mixed traffic. In addition, some regulation is needed to standardize technologies across manufacturers, to ensure that vehicles communicate accurately with other vehicles and with the enabling infrastructure.

---

[12] *Casual carpooling* refers to carpools formed at pickup locations, at which a driver pulls up and picks up one or two passengers. Such carpooling mechanisms have formed in a few large metropolitan areas with HOV lanes.

[13] This discussion refers to surface-roadway vehicles. Driverless vehicles in other modes, such as driverless trains and unmanned aerial vehicles, have been in use for decades.

Change Liability Laws to Enable Use of Driverless Vehicles

In addition to perfecting the technology, the main obstacles to widespread adoption of driverless vehicles are related to liability. As vehicles increasingly adopt technologies that bypass the driver's decisionmaking processes, the liability for a crash shifts from the driver to the manufacturer. Although autonomous technologies can increase overall safety by reducing driver error, which would suggest that technologies might be adopted quickly, concerns about liability might discourage manufacturers from incorporating such technologies for fear of lawsuits (Kalra, Anderson, and Wachs, 2009). Creating a comprehensive regulatory regime at the federal level, to preempt having conflicting state laws, might provide an environment that allows faster adoption of driverless vehicles.

CHAPTER FIVE

Applying Energy Services Analysis to Other Contexts

In the previous two chapters, we used ESA to estimate potential reductions in energy use and GHG emissions from delivering news using e-readers rather than newspapers and from providing mobility through vehicle sharing rather than owning private motor vehicles. This chapter presents several brief examples of other ways of fulfilling the needs laid out by Costanza et al. (2007) (e.g., subsistence, security, spirituality). For each example, we identify the need, describe the current approach to satisfying the need, discuss one or more alternative approaches to satisfying the need, and discuss the potential benefits in terms of reductions in energy use and GHG emissions that could result from satisfying the need in an alternative fashion.

The alternatives discussed in the previous chapters are commercially available. In this chapter, some of the alternatives suggested are not currently commercially available. Although applying ESA to such alternatives is possible, the level of analysis will naturally be limited.

Food

Food is necessary for subsistence. Because no realistic alternative service can replace food, ESA needs to focus on alternative ways of growing, processing, transporting, and preparing food, all of which use energy, directly and indirectly.

One recent approach to procuring food has focused on procuring locally grown food. However, it is not clear whether procuring locally grown products actually saves energy. ESA can be used to evaluate various ways of procuring similar foods from various locations, identifying the most energy-efficient way to procure a specific food. Location is particularly important with respect to food because certain climates and land types lend themselves to specific types of food production more readily than others. A New Zealand study finds that less energy was consumed when some foods were grown in New Zealand and shipped to the United Kingdom than when the food was produced in the United Kingdom. Certain crops and animals in New Zealand required so much less feed or fertilizer when produced in New Zealand that this difference more than compensated for the energy used to transport it to the United Kingdom (Saunders, Barber, and Taylor, 2006).

ESA can also be used to evaluate alternative means of procuring a set amount of protein, for example, comparing the energy involved in producing and procuring plant and animal proteins: the energy needed to produce the food (including the energy needed to grow feed for animals), slaughter or harvest the food, process it, transport it to consumers, and cook it.

Food preparation accounts for a considerable share of energy used in the home. ESA can be used to measure differences in improving energy efficiency in food preparation by improving the efficiency of home appliances or by buying prepared food outside the home. In some instances, less energy might be expended purchasing food from restaurant kitchens than cooking at home.

Some modes of cooking use more energy than others. For example, boiling a large pot of water uses more energy than stir-frying or microwaving. New models of food preparation may reduce energy use, such as increased use of local or decentralized food-preparation services (e.g., neighborhood food preparers that focus on the creation of prepared meals that can be reheated at home), or home chef services, in which a household's weekly meals might be prepared in the course of one day and refrigerated.

Part of the analysis could also look at the structure of American diets to determine whether shifts in the categories of foods consumed might lead to decreased energy use. More energy may be entailed in producing some types of processed foods than others. ESA is a means of measuring these differences, making it possible to measure the relative energy use of in-home preparation versus large-scale processing.

Clothing

Clothing provides protection from heat and cold and is a means of expressing one's creativity and identity. ESA provides a means of evaluating the implications for energy consumption of alternative approaches to styles and manufacturing clothing, including such issues as alternative materials, production techniques, transportation, consumption, and disposal.

Clothing is manufactured from a large variety of materials that may be grown, synthesized, or transported using more or less energy. ESA highlights energy trade-offs in terms of material, longevity of the product, and care. For example, a material may be more energy intensive to fabricate; however, if the clothes made from the material last longer, ESA would reveal the true energy balance. ESA reveals the energy consumption involved in cleaning and maintaining a wardrobe, identifying potential reductions in energy use from washing and ironing through the use of fabrics and coatings that effectively resist stains and wrinkles. ESA can also be used to incorporate the energy implications of using fabrics that can be laundered rather than dry cleaned.

Assessing potential reductions in energy use in the clothing sector may also entail evaluating the implications of networks that are more efficient for redistribution (such as greater use of consignment and thrift stores), so that used clothing can be passed on to other consumers, as well as expanded opportunities for renting clothing that will be worn a limited number of times, such as wedding gowns.

ESA can also identify potential energy savings if people were to dress more warmly in winter, making it possible to reduce temperatures at work (or home) and dress more coolly in summer, so that room temperatures could be set at higher levels, reducing the need for air conditioning.

Health Care

Residents of developed countries spend a rising share of income on health care in the quest to prolong and improve the quality of their lives. People use a large array of practices, technologies, and treatments to stay healthy or treat illnesses. ESA can be used to evaluate the implications of the various alternatives to delivering and receiving health care in terms of consuming energy.

ESA could be employed to evaluate opportunities for reducing energy use associated with hospitals and other health care facilities. ESA could be used to evaluate the implications for energy consumption of delivering more health care in the home rather than in specialized facilities. The energy consumption associated with constructing, heating, cooling, and supplying electricity to hospitals and clinics would need to be balanced against the increased use of energy in the form of visits by medical personnel to the home or operation of medical equipment in the home. Remote access to medical personnel, such as "telecaregiving," whereby patients may be monitored in their homes by camera, may save energy involved in transportation through fewer visits to patients' homes (Ludden, 2010).

ESA may also be used to evaluate new alternatives for specific treatments that might reduce energy consumption. These may include drugs that can be customized to individuals, bio- and nanomaterials that can be used to repair or regrow damaged tissue or bone, "shape memory" materials that could be used to fabricate medical instruments that reduce the need for invasive procedures, and robotic devices that can assist in surgery (Silberglitt et al., 2006). ESA can be used to compare energy use involved in these innovative treatments with use by their current counterparts.

Waste Disposal

Waste handling and treatment are critical services for both households and firms, necessary for subsistence and for maintaining quality of life. Most municipal waste has conventionally been disposed of in landfills, where it slowly decomposes over time. In the past few decades, growing recognition of potential environmental hazards from landfills has spurred the development of alternative waste-management methods, such as recycling and incineration.

ESA provides a useful tool to evaluate the energy implications of landfills compared to other forms of waste disposal. An analysis would incorporate assessments of the heat generated from the incineration of waste material used to replace other energy sources. An analysis of landfills would assess the energy involved in the construction of liner systems, collection of the gas generated by the landfill, prevention of leaching or collection of wastes that leach from the landfill, and ongoing maintenance. Assessing recycling would require estimating the energy used in sorting, direct energy use in converting the waste into new materials, and transportation of the new materials to facilities where they will be used. ESA would balance these energy uses against the energy savings from replacing newly manufactured materials with recycled products. The embodied energy in the capital equipment used in the alternative approaches to disposal would be assessed by ESA. For example, an analysis of waste-management techniques by a research group in Sweden compares the energy use of landfills, recycling, and incineration with heat recovery and finds that recycling generally had the lowest energy use. However, the study finds that, under certain assumptions about what products the recycled materials would replace, incineration had a lower energy cost than recycling (Finnveden et al., 2000). This case illustrates the importance of the specific operations for determining actual energy consumption.

Relative Importance of Energy Savings, by Service Area

Each of the examples described in this chapter suggests ways in which a particular need might be met in a manner that uses less energy. Given the large number of possible ways to reduce energy use, it is useful to focus on services that account for a large share of energy use.

Table 5.1 shows estimated energy use for some consumer activities in the United States. The data are based on a study of direct and indirect household energy use by Bin and Dowlatabadi

Table 5.1
U.S. Residential Energy Consumption

Energy Type	Consumer Activity	Energy Use (EJ)
Direct		28.3
Home energy		10.9
	Space heating	5.5
	Other appliances and lighting	2.4
	Water heating	2.0
	Refrigeration	0.5
	Air conditioning	0.4
Personal travel		17.4
	Short distance by automobile or truck	12.7
	Long distance by airplane	2.2
	Long distance by automobile	1.9
	Long distance by other mode	0.3
	Short distance by other mode	0.3
Indirect		56.4
	Housing operation (shelter, utilities, household operations and furnishings)	25.6
	Transportation operation (vehicle purchase, gasoline, motor oil, other vehicle expenses)	17.4
	Food and beverages (at and away from home)	6.4
	Apparel and services	2.6
	Other (e.g., education, tobacco)	1.6
	Personal insurance and pensions	1.1
	Entertainment (e.g., fees and admissions, electronic and sports equipment, reading material)	1.1
	Health care (health insurance, medical services, drugs, medical supplies)	0.5
Total		84.6

SOURCE: Bin and Dowlatabadi, 2005, Table 3.
NOTE: EJ = exajoule (1,018 joules).

(2005), which drew data on direct energy use from EIA's Residential Energy Consumption Survey database, DOE's Transportation Energy Data Book, and the Department of Transportation's American Travel Survey; data on indirect energy use are estimates by Bin and Dowlatabadi based on the Department of Commerce's Consumer Expenditure Survey and the Environmental Input-Output Life Cycle Analysis model developed by researchers at Carnegie Mellon University.

Table 5.1 demonstrates that focusing only on direct energy use, the conventional energy accounting approach, fails to account for the nearly two-thirds of annual household energy use that is consumed indirectly. The top five categories of energy use, in descending order, are housing operation (including shelter and utilities), transportation operation (including vehicle purchase, gasoline and motor oil, and other vehicle expenses), short-distance trips by automobile and truck, food and beverages, and space heating. These five categories together account for 80 percent of household energy consumption. This list identifies those services that currently consume the greatest shares of household energy use and where alternative, less energy-intensive ways to provide these services may provide the greatest potential for energy savings.

CHAPTER SIX

Conclusion

We conclude with two observations about the utility of ESA as a framework for seeking unconventional ways to reduce energy use and GHG emissions. Table 6.1 summarizes the differences in per capita and aggregate decreases for both of our examples, as well as our assumptions about the diffusion of the alternative way of providing the service.

The first observation is that there are large differences in emission reductions depending whether the reduction is measured as per capita or aggregate, and as percentage or absolute. For the delivery of written news, the percentage reduction estimated in this summary is as much as 89 percent, both on a per-user basis and in aggregate, assuming that all users switch from newsprint to reading electronically. The absolute reductions could be as high as 84.1 kg of CO_2e per user, or 3.87 million metric tons overall. Reducing GHG emissions by similar amounts by improving the efficiency of publishing paper newspapers is infeasible. Even in the future, technical improvements in the efficiency of producing paper or printing newspapers could not generate such large reductions in energy use or GHG emissions. Although the total emissions of 4.35 million metric tons of GHG emissions from the newspaper industry are not a large total of U.S. GHG emissions, they represent a substantial reduction for one sector.

Table 6.1
Comparison of Estimated Greenhouse Gas Emission Reductions for News Delivery and Personal Mobility, Per Capita and Total

Factor	Written News	Personal Mobility (maximum adoption case)
Reductions per user		
Current annual GHG emissions (kg CO_2e)	94.7	2,380
Maximum potential annual GHG emissions avoided with alternative service provision (kg CO_2e)	84.3	893
Maximum percentage reduction	89	37
Total reductions		
Current annual GHG emissions for sector (millions of metric tons)	4.35	1,071
Assumed percentage of users who adopt alternative service provision	100	12.5
Maximum potential annual GHG emissions avoided with alternative service provision (kg CO_2e)	3.87	18.1
Maximum percentage reduction	89	1.7

Vehicle-sharing programs also generate substantial per-user reductions in GHG emissions, about 37 percent per user. However, only 0.27 percent of U.S. drivers currently participate in vehicle-sharing programs; maximum potential participation has been estimated at 12.5 percent. Participation is constrained by the availability of public transportation, the availability of vehicle-sharing services, and ease of use. Moreover, the profile of those who join vehicle-sharing organizations is quite different from those who do not: Members of vehicle-sharing organizations drive far fewer miles annually than the average U.S. driver. In the unlikely event that maximum potential participation is achieved, the corresponding reduction in total U.S. GHG emissions would be on the order of 18.1 million metric tons. Because potential likely members drive so much less than the national average, this reduction in emissions would be equivalent to only 1.7 percent of total GHG emissions (1,071 million metric tons) from light-duty vehicles. Thus, although the relative size of reductions may be large, overall reductions in the context of total U.S. emissions of GHGs would likely be relatively small.

The point is not that we should focus all of our efforts on a few key sectors. GHG emissions and energy consumption are inherent in almost all of our daily activities, and the search for a "silver bullet" that will reduce GHGs substantially and easily is a misleading one. However, it seems reasonable that we should focus our collective efforts on services in which higher reductions are likely, and a cross-service comparison can yield insights into where those efforts might be most productively focused.

A second observation is that our estimates for potential reductions in energy consumption and GHG emissions are very heavily dependent on assumptions—in some cases, assumptions for which good data are not available. Three assumptions about newspapers lead to fairly large variations in estimates of current GHG emissions: the share of recycled paper, the emission profile of the energy needed for production, and the share of newspapers that are recycled. If these three assumptions are additive, actual emissions could be as much as 65 percent lower than what we calculated. In this case, existing data support our assumptions, and we believe them to be reasonably accurate.

However, our results about tablet computers and e-readers, and the comparison with newspapers, are equally dependent on assumptions; in this case, data are far less available. For example, no leading e-reader manufacturer has made public the emission profile of its device. As a result, our analysis is based on data that, although only a few years old, most likely do not reflect the rapid technological changes in the consumer electronics industry. (Indeed, while writing this report, we changed our models to incorporate new information on devices that did not exist when we began this analysis.) Our assumptions are even more uncertain on such issues as the effective life of an electronic device; the amount of time users spend reading newspapers on an e-reader as opposed to magazines, books, or other media; and the number of print subscriptions or books replaced by one device. Changes in any of these assumptions would change our results, possibly by fairly large percentages.

Assumptions also affect the magnitude of potential reductions from vehicle sharing. One key assumption is that vehicle-sharing members drive far less on average than typical American drivers. Although this is based on existing data, if vehicle sharing were to grow substantially, it would per force attract members who drive somewhat more VMT. So the per capita and aggregate decreases would become larger. Aggregate decreases also depend on the percentage of drivers who join vehicle-sharing organizations; changes in these assumptions would also have substantial impacts.

These limitations are not exclusive to ESA, which we consider a valuable framework to use in assessing areas in which energy use and GHG emissions might be reduced. ESA provides a new lens through which to view the energy used in a variety of areas to fulfill human wants and needs. We expect this and other such applications, some of which we suggested in the previous chapter, to lead to fresh ideas for reducing our energy use and GHG emissions.

Energy Consumption Associated with E-Readers

Table A.1 shows the calculations from million joules to kilograms of CO_2e from the data provided in Moberg et al. (2009).

Table A.1
Conversion of Million Joules of Energy Consumption from an iRex Iliad E-Reader to Kilograms of Carbon Dioxide Equivalent Emissions

Emission	Energy Consumption (MJ)	GHG Emissions (kg of CO_2e)
From indirect energy use		
Manufacturing	191	11.0
Distribution	1	0.04
Disposal	−1	−0.06
Subtotal	191	11.0
From device and ICT network energy use	237	13.7
Total	428	24.7

SOURCES: Moberg et al. (2009); EPA, 2010.

NOTE: Indirect energy-use data come from Moberg et al. (2009, Appendix 4.8). Device and ICT network energy-use data are from Moberg et al. (2009, p. 3). The figure provided is 14–25 watt-hours (Wh) per day, which is based on research by Taylor and Koomey (2008). To make this comparable with other estimates, we averaged the 14–25 range, multiplied the daily energy use by 365 days per year, divided by 1,000 to change from watt-hours to kilowatt-hours, multiplied by 3.6 MJ/kWh, multiplied by 3.079 to convert from site to source energy, and multiplied this annual figure by 3 years. We generated estimates of GHG emissions by using the U.S. national average for CO_2 emissions per kilowatt-hour of net power, which we have converted to 0.0577 kg/MJ (see EPA, 2010, footnote 5).

References

Abdel-Razzaq, Laurén, "RelayRides Redesigns Car Sharing; Neighbors Borrow Cars from Each Other," *Automotive News*, May 9, 2011.

Abell, John C., "And the Most Popular Way to Read an E-Book Is . . . ," *Wired*, November 8, 2010. As of February 20, 2012:
http://www.wired.com/epicenter/2010/11/and-the-most-popular-way-to-read-an-e-book-is/

American Automobile Association, "Your Driving Costs," undated. As of November 5, 2010:
http://www.aaaexchange.com/Main/Default.asp?CategoryID=16&SubCategoryID=76&ContentID=353

Apple, *iPad Environmental Report*, 2010. As of April 22, 2010:
http://images.apple.com/environment/reports/docs/iPad_Environmental_Report.pdf

———, *iPad 2 Environmental Report*, 2011. As of February 6, 2012:
http://www.apple.com/environment/reports/docs/iPad_2_Environmental_Report.pdf

Arnott, Sarah, "Ahmed al-Jaber: Oil-Rich Abu Dhabi Pins Its Hopes on Dreams of a Green Future," *Independent*, February 4, 2010. As of February 6, 2012:
http://www.independent.co.uk/news/people/profiles/
ahmed-aljaber-oilrich-abu-dhabi-pins-its-hopes-on-dreams-of-a-green-future-1888901.html

Bandivadekar, Anup, Kristian Bodek, Lynette Cheah, Christopher Evans, Tiffany Groode, John Heywood, Emmanuel Kasseris, Matthew Kromer, and Malcolm Weiss, *On the Road in 2035: Reducing Transportation's Petroleum Consumption and GHG Emissions*, Cambridge, Mass.: Massachusetts Institute of Technology, July 2008.

Barone, Jennifer, Amber Fields, Karen Rowan, and Jessica Ruvinsky, "How Big Is DISCOVER's Carbon Footprint," *DISCOVER Magazine*, May 2008. As of April 25, 2010:
http://discovermagazine.com/2008/may/21-how-big-is-discover.s-carbon-footprint

Beer, Jeroen de, *Potential for Industrial Energy-Efficiency Improvement in the Long Term*, Dordrecht, Netherlands: Kluwer Academic, 2000.

Belson, Ken, "Baby, You Can Rent My Car," *New York Times*, September 10, 2010. As of February 6, 2012:
http://www.nytimes.com/2010/09/12/automobiles/12RELAY.html

Bhatia, Pankaj, and Janet Ranganathan, *The Greenhouse Gas Protocol: A Corporate Accounting and Reporting Standard*, rev. ed., Washington, D.C., World Resources Institute, March 2004. As of October 5, 2011:
http://www.wri.org/publication/
greenhouse-gas-protocol-corporate-accounting-and-reporting-standard-revised-edition

Bieszczat, Alice, and Joseph Schwieterman, *Are Taxes on Carsharing Too High? A Review of the Public Benefits and Tax Burden of an Expanding Transportation Sector*, Chicago, Ill.: Chaddick Institute for Metropolitan Development, DePaul University, June 28, 2011. As of February 6, 2012:
http://las.depaul.edu/chaddick/docs/Docs/DePaul_University_Study_on_Taxation_of_C.pdf

Bin, Shui, and Hadi Dowlatabadi, "Consumer Lifestyle Approach to US Energy Use and the Related CO_2 Emissions," *Energy Policy*, Vol. 33, No. 2, January 2005, pp. 197–208.

Binns, Shannon, Tyson Miller, Nicole Rycroft, and Shiloh Bouvette, *A Brighter Shade of Green: Opportunities for Newspapers in the New Era of Consumer Environmentalism—2008 Industry Briefing Report*, Green Press Initiative and Markets Initiative, 2008. As of April 22:
2010:
http://www.greenpressinitiative.org/documents/newspaperreport.pdf

Blottnitz, Harro von, and Mary Ann Curran, "A Review of Assessments Conducted on Bio-Ethanol as a Transportation Fuel from a Net Energy, Greenhouse Gas, and Environmental Life Cycle Perspective," *Journal of Cleaner Production*, Vol. 15, 2007, pp. 607–619.

Bonner, Jessie L., "Car Sharing Steers Congestion Away from University Campuses," *Washington Post*, October 10, 2010. As of February 20, 2012:
http://www.washingtonpost.com/wp-dyn/content/article/2010/10/09/AR2010100900077.html

Borealis Centre for Environment and Trade Research, "Environmental Trends and Climate Impacts: Findings from the U.S. Book Industry," Book Industry Study Group and Green Press Initiative, 2008. As of April 22, 2010:
http://www.greenpressinitiative.org/documents/trends_summary.pdf

Borkowski, Liz, and Melissa Kelley, *Turning the Page: Environmental Impacts of the Magazine Industry and Recommendations for Improvement*, Washington, D.C.: PAPER Project, May 2001. As of April 22, 2010:
http://www.greenamericatoday.org/pdf/whitepapermagazines.pdf

Cabanatuan, Michael, "State Bill Offers Twist to Expand Car Sharing," *San Francisco Chronicle*, April 29, 2010. As of February 6, 2012:
http://www.sfgate.com/cgi-bin/article.cgi?f=/c/a/2010/04/28/BA1B1D6DR5.DTL

California Assembly, Assembly Bill 1871, filed with Secretary of State September 29, 2010. As of February 9, 2012:
http://www.leginfo.ca.gov/pub/09-10/bill/asm/ab_1851-1900/ab_1871_bill_20100929_chaptered.html

California State Association of Counties, "South Bay Cities of L.A. County Partner with Enterprise Fleet Management in Unique Pilot Car Project," *CASC Bulletin*, August 6, 2010. As of October 12, 2011:
http://bulletin.counties.org/sec.aspx?id=D0F72826EEA04325841588FCE718A674

Car2go, "Featured Apps," undated. As of February 20, 2012:
http://www.car2go.com/apps/en/overview/

Carbon Trust, *Carbon Footprints in the Supply Chain: The Next Step for Business*, London, CTC616, November 20, 2006.

CarSharing.net, "Where Can I Find Car Sharing?" undated. As of February 6, 2012:
http://www.carsharing.net/where.html

Celsor, Christine, and Adam Millard-Ball, "Where Does Carsharing Work? Using Geographic Information Systems to Assess Market Potential," *Transportation Research Record*, Vol. 1992, 2007, pp. 61–69.

Chan, Nelson D., and Susan A. Shaheen, "Ridesharing in North America: Past, Present, and Future," *Transport Reviews*, Vol. 32, No. 1, 2011, pp. 93–112.

Chase, Katie Johnston, "Zipcar, Still Awaiting Profit, Plans to Go Public," *Boston Globe*, June 2, 2010. As of February 7, 2012:
http://www.boston.com/business/articles/2010/06/02/zipcar_still_awaiting_profit_plans_to_go_public/

City CarShare, "Newsletter," San Francisco, Calif., June 2009.

Code of Federal Regulations, Title 49, Transportation, Chapter V, National Highway Traffic Safety Administration, Department of Transportation, Part 571, Federal motor vehicle safety standards, Standard 500, Low-speed vehicles. As of February 6, 2012:
http://edocket.access.gpo.gov/cfr_2003/octqtr/pdf/49cfr571.500.pdf

Committee for the Study on the Relationships Among Development Patterns, Vehicle Miles Traveled, and Energy Consumption, National Research Council, *Driving and the Built Environment: The Effects of Compact Development on Motorized Travel, Energy Use, and CO$_2$ Emissions*, Washington, D.C.: Transportation Research Board, Special Report 298, 2009. As of February 6, 2012:
http://www.nap.edu/catalog.php?record_id=12747

Costanza, Robert, Brendan Fisher, Saleem Ali, Caroline Beer, Lynne Bond, Roelof Boumans, Nicholas L. Danigelis, Jennifer Dickinson, Carolyn Elliott, Joshua Farley, Diane Elliott Gayer, Linda MacDonald Glenn, Thomas Hudspeth, Dennis Mahoney, Laurence McCahill, Barbara McIntosh, Brian Reed, S. Abu Turab Rizvi, Donna M. Rizzo, Thomas Simpatico, and Robert Snapp, "Quality of Life: An Approach Integrating Opportunities, Human Needs, and Subjective Well-Being," *Ecological Economics*, Vol. 61, 2007, pp. 267–276.

Curran, Mary Ann, and Philippa Notten, *Summary of Global Life Cycle Inventory Data Resources*, prepared for Task Force 1: Database Registry Society for Environmental Toxicology and Chemistry/United Nations Environment Programme Life Cycle Initiative, May 2006. As of February 6, 2012:
http://www.epa.gov/nrmrl/lcaccess/pdfs/summary_of_global_lci_data_resources.pdf

Dancing Rabbit Ecovillage, "DRVC," undated. As of February 20, 2012:
http://www.dancingrabbit.org/drvc/

Davis, Stacy, Susan Diegel, and Robert Boundy, *Transportation Energy Data Book: Edition 28*, Oak Ridge, Tenn.: Oak Ridge National Laboratory, ORNL-6984, 2009.

———, *Transportation Energy Data Book: Edition 30*, Oak Ridge, Tenn.: Oak Ridge National Laboratory, ORNL-6986, June 2011.

Dodson, Sean, "Technology: Welcome to the Transport of Tomorrow: First Mooted over a Century Ago, Personal Rapid Transit Systems Might Soon Be Running Through Our Cities," *Guardian*, October 11, 2007.

Douma, Frank, and James Andrew, "Developing a Model for Car Sharing Potential in Twin Cities Neighborhoods," in Transportation Research Board, *TRB 85th Annual Meeting Compendium of Papers*, Paper 06-2449, 2006.

DriveMyCar Rentals, "About Us," undated. As of November 5, 2010:
http://www.drivemycarrentals.com.au/about.asp

Ecola, Liisa, Scott Hassell, Michael Toman, and Martin Wachs, *Integrating U.S. Climate, Energy, and Transportation Policies: Proceedings of Three Workshops*, Santa Monica, Calif.: RAND Corporation, CF-256-MCCORF, 2009. As of February 7, 2012:
http://www.rand.org/pubs/conf_proceedings/CF256.html

EIA—*See* U.S. Energy Information Administration.

EPA—*See* U.S. Environmental Protection Agency.

Fazzalaro, James J., *Neighborhood Electric Vehicles*, Hartford, Conn.: Connecticut General Assembly, Office of Legislative Research, 2008-R-0479, September 22, 2008.

Federal Highway Administration, licensed drivers, by age (Table DL-22), *Highway Statistics 2010*, Washington, D.C., September 2011. As of February 6, 2012:
http://www.fhwa.dot.gov/policyinformation/statistics/2010/dl22.cfm

Finnveden, Göran, Jessica Johansson, Per Lind, and Åsa Moberg, *Life Cycle Assessments of Energy from Solid Waste*, Stockholm: Stockholm University, August 2000. As of February 6, 2012:
http://www.imamu.edu.sa/topics/IT/IT%206/
Life%20Cycle%20Assessments%20of%20Energy%20from%20Solid%20Waste.pdf

Firnkorn, Jörg, and Martin Müller, "What Will Be the Environmental Effects of New Free-Floating Car-Sharing Systems? The Case of car2go in Ulm," *Ecological Economics*, Vol. 70, No. 8, June 2011, pp. 1519–1528.

Gower, Stith T., Ann McKeon-Ruediger, Annabeth Reitter, Michael Bradley, David J. Refkin, Timothy Tollefson, Fred J. Souba Jr., Amy Taup, Lynn Embury-Williams, Steven Schiavone, James Weinbauer, Anthony C. Janetos, and Ron Jarvis, "Following the Paper Trail: The Impact of Magazine and Dimensional Lumber Production on Greenhouse Gas Emissions—A Case Study," Washington, D.C.: H. John Heinz III Center for Science, Economics, and the Environment, 2006.

Haas, Reinhard, Nebojsa Nakicenovic, Amela Ajanovic, Thomas Faber, Lukas Kranzl, Andreas Müller, and Gustav Resch, "Towards Sustainability of Energy Systems: A Primer on How to Apply the Concept of Energy Services to Identify Necessary Trends and Policies," *Energy Policy*, Vol. 36, No. 11, November 2008, pp. 4012–4021.

Haefeli, Ueli, Daniel Matti, Christoph Schreyer, and Markus Maibach, *Evaluation Car-Sharing*, Bern, Switzerland: Bundesamt für Energie, September 2006.

Hamilton, Tyler, "Taking Car-Sharing to the Max," (Toronto) *Star*, February 20, 2010. As of February 7, 2012:
http://www.thestar.com/business/article/768533

Harms, Sylvia, and Bernard Truffer, *The Emergence of a Nation-Wide Carsharing Co-Operative in Switzerland*, Switzerland: Swiss Federal Institute of Aquatic Science and Technology, 1998.

Hartwig, Stephan, and Michael Buchmann, *Empty Seats Traveling: Next-Generation Ridesharing and Its Potential to Mitigate Traffic- and Emission Problems in the 21st Century*, Bochum, Germany: Nokia Research Center, NRC-TR-2007-003, February 14, 2007. As of February 7, 2012:
http://research.nokia.com/files/tr/NRC-TR-2007-003.pdf

Higginbotham, Brian, *Carsharing: Increasing Rural Transportation Options in the Great Central Valley*, Modesto, Calif.: Great Valley Center, Fall 2000.

Hillsman, Edward L., "Book Review of 'Spatial Energy Analysis,' Edited by Lars Lundqvist, Lars-Goran Mattsson, and Erik A. Eriksson," *Journal of Regional Science*, Vol. 31, No. 2, 1991, pp. 233–235.

Innovative Mobility Research, "Carsharing," last updated February 6, 2012, referenced October 5, 2011. As of February 7, 2012:
http://www.innovativemobility.org/carsharing/index.shtml

Insley, Jill, "WhipCar Launches Neighbour-to-Neighbour Rental Scheme," *Guardian*, April 22, 2010.

International Organization for Standardization, *Environmental Management—Life Cycle Assessment—Principles and Framework*, ISO 14040:2006, July 17, 2010a.

———, *Environmental Management—Life Cycle Assessment—Requirements and Guidelines*, ISO 14044:2006, July 17, 2010b.

ISO—*See* International Organization for Standardization.

Jochem, Eberhard, Anthony Adegbulugbe, Bernard Aebischer, Somnath Bhattacharjee, Inna Gritsevich, Gilberto Jannuzzi, Tamas Jaszay, Bidyut Baran Saha, Ernst Worrell, and Zhou Fengqi, "Chapter 6: Energy End-Use Efficiency," *World Energy Assessment: Energy and the Challenge of Sustainability*, New York: United Nations Development Programme, 2000, pp. 173–218.

Kahn, Debra, "Levi Strauss, Goodwill Aim to Lessen Your Laundry Blues," *ClimateWire*, October 22, 2009.

Kalra, Nidhi, James M. Anderson, and Martin Wachs, *Liability and Regulation of Autonomous Vehicle Technologies*, Berkeley, Calif.: California PATH Program, Institute of Transportation Studies, University of California at Berkeley, California PATH Research Report UCB-ITS-PRR-2009-28, April 2009. As of February 7, 2012:
http://www.rand.org/pubs/external_publications/EP20090427.html

Kessler, David S., *Computer-Aided Scheduling and Dispatch in Demand-Responsive Transit Services: A Synthesis of Transit Practice*, Washington, D.C.: Transportation Research Board, Transit Cooperative Research Program Synthesis 57, 2004. As of February 7, 2012:
http://www.trb.org/Main/Public/Blurbs/155357.aspx

Kinsella, Susan, Gerard Gleason, Victoria Mills, Nicole Rycroft, Jim Ford, Kelly Sheehan, and Joshua Martin, *The State of the Paper Industry: Monitoring the Indicators of Environmental Performance*, Environmental Paper Network, 2007.

Kirshner, Dan, "Pilot Tests of Dynamic Ridesharing," updated January 23, 2007. As of February 7, 2012:
http://www.ridenow.org/ridenow_summary.html

Kozak, Greg, *Printed Scholarly Books and E-Book Reading Devices: A Comparative Life Cycle Assessment of Two Book Options*, Ann Arbor, Mich.: Center for Sustainable Systems, University of Michigan, Report CSS03-04, August 24, 2003. As of April 22, 2010:
http://css.snre.umich.edu/css_doc/CSS03-04.pdf

Litman, Todd, *Evaluating Carsharing Benefits*, Victoria, B.C.: Victoria Transport Policy Institute, December 7, 1999. As of February 7, 2012:
http://www.vtpi.org/carshare.pdf

Loose, Willi, Mario Mohr, and Claudia Nobis, "Assessment of the Future Development of Car Sharing in Germany and Related Opportunities," *Transport Reviews*, Vol. 26, No. 3, 2006.

Lu, S., *Vehicle Survivability and Travel Mileage Schedules*, Washington, D.C.: National Highway Traffic Safety Administration, National Center for Statistics and Analysis, DOT HS 809 952, January 2006. As of October 6, 2011:
http://www-nrd.nhtsa.dot.gov/Pubs/809952.PDF

Ludden, Jennifer, "Wired Homes Keep Tabs on Aging Parents," National Public Radio, August 24, 2010. As of November 5, 2010:
http://www.npr.org/templates/story/story.php?storyId=129104664

Martin, Elliot W., and Susan A. Shaheen, *Greenhouse Gas Emission Impacts of Carsharing in North America*, San Jose, Calif.: Mineta Transportation Institute, San Jose State University, MTI Report 09-11, June 2010.

———, "Greenhouse Gas Emission Impacts of Carsharing in North America," *IEEE Transactions on Intelligent Transportation Systems,* Vol. 12, No. 4, December 2011, pp. 1074–1086.

Martin, Elliot, Susan A. Shaheen, and Jeffrey R. Lidicker, "Carsharing's Impact on Household Vehicle Holdings: Results from a North American Shared-Use Vehicle Survey," *Annual Meeting of the Transportation Research Board*, Washington, D.C., 2010.

Massey, Ray, "Car Firms Pin Hopes on Pay-as-You-Drive as Recession Makes Buying Vehicles Too Expensive," (London) *Daily Mail*, May 8, 2010. As of February 7, 2012:
http://www.dailymail.co.uk/news/article-1274879/
Car-firms-pin-hopes-pay-drive-recession-makes-buying-vehicles-expensive.html

Millard-Ball, Adam, Gail Murray, Jessica ter Schure, Christine Fox, and Jon Burkhardt, *Car-Sharing: Where and How It Succeeds*, Washington, D.C.: Transportation Research Board, Transit Cooperative Research Program Report 108, 2005.

Moberg, Åsa, Martin Johansson, Göran Finnveden, and Alex Jonsson, *Screening Environmental Life Cycle Assessment of Printed, Web Based and Tablet E-Paper Newspaper*, 2nd ed., Stockholm: KTH Centre for Sustainable Communications, TRITA-SUS Report 2007:1, 2009. As of April 22, 2010:
http://www.sustainablecommunications.org/wp-content/publications/2009-Report-Screening-newspaper.pdf

Mobility CarSharing, "History, Figures," 2010.

Moffat, Amy, director, research and communications, Great Valley Center, email discussion with the author, June 8, 2010.

Moore, Charles W., "Requiem for the Crown Vic," *Telegraph Journal* (New Brunswick, Canada), September 29, 2011.

Morrison, Peter A. "Gauging Future Prospects for a Neighborhood Vehicle: Where Demographic Analysis Fits In," *Australian Population Association's 11th Biennial Conference*, Sydney, October 2–4, 2002. As of February 7, 2012:
http://www.apa.org.au/upload/2002-4A_Morrison.pdf

NAA—*See* Newspaper Association of America.

National Household Travel Survey, home page, undated. As of November 8, 2010:
http://nhts.ornl.gov/

National Renewable Energy Laboratory, "U.S. Life Cycle Inventory Database," last updated January 5, 2012. As of February 20, 2012:
http://www.nrel.gov/lci/

Newspaper Association of America, "Newspaper Circulation Volume," undated. As of October 5, 2011:
http://www.naa.org/Trends-and-Numbers/Circulation/Newspaper-Circulation-Volume.aspx

Nors, Minna, Tiina Pajula, and Hanna Pihkola, "Calculating the Carbon Footprints of a Finnish Newspaper and Magazine from Cradle to Grave," in Heli Koukkari and Minna Nors, eds., *Life Cycle Assessment of Products and Technologies: LCA Symposium—VTT, Espoo, Finland, 6 October 2009*, Espoo, Finland: VTT, Symposium 262, 2009, pp. 55–65. As of February 7, 2012:
http://www.vtt.fi/inf/pdf/symposiums/2009/S262.pdf

NREL—*See* National Renewable Energy Laboratory.

Oregon Legislative Assembly, House Bill 3149, enrolled, 2011. As of February 9, 2012:
http://www.leg.state.or.us/11reg/measpdf/hb3100.dir/hb3149.en.pdf

Ortega, Juan, *Car Sharing in the United States: Helping People Transition from Welfare to Work and Improving the Quality of Life of Low-Income Families*, Washington, D.C.: Community Transportation Association of America, c. 2005. As of February 7, 2012:
http://web1.ctaa.org/webmodules/webarticles/articlefiles/carsharing_report_final.pdf

Ovide, Shira, "U.S. Newspaper Circulation Falls," *Wall Street Journal*, October 27, 2009.

Park, Hi-Chun, and Eunnyeong Heo, "The Direct and Indirect Household Energy Requirements in the Republic of Korea from 1980 to 2000: An Input-Output Analysis," *Energy Policy*, Vol. 35, No. 5, May 2007, pp. 2839–2851.

Perez-Pena, Richard, "U.S. Newspaper Circulation Falls 10%," *New York Times*, October 26, 2009. As of April 15, 2010:
http://www.nytimes.com/2009/10/27/business/media/27audit.html

Peter Muheim and Partner, "CarSharing: The Key to Combined Mobility," Bern: Swiss Federal Office of Energy, September 1998.

Pew Research Center for the People and the Press, "Press Accuracy Rating Hits Two Decade Low," September 13, 2009. As of February 20, 2012:
http://www.people-press.org/2009/09/13/press-accuracy-rating-hits-two-decade-low/

Pihkola, Hanna, Minna Nors, Marjukka Kujanpää, Tuomas Helin, Merja Kariniemi, Tiina Pajula, Helena Dahlbo, and Sirkka Koskela, *Carbon Footprint and Environmental Impacts of Print Products from Cradle to Grave: Results from the LEADER Project (Part 1)*, Espoo, Finland: VTT Tiedotteita, Research Notes 2560, 2010. As of February 7, 2012:
http://www.vtt.fi/inf/pdf/tiedotteet/2010/T2560.pdf

Pisarski, Alan E., *Commuting in America III: The Third National Report on Commuting Patterns and Trends*, Washington, D.C.: Transportation Research Board, National Cooperative Highway Research Program Report 550/Transit Cooperative Research Program Report 110, 2006. As of February 7, 2012:
http://onlinepubs.trb.org/onlinepubs/nchrp/ciaiii.pdf

Plambeck, Joseph, "Newspaper Circulation Falls Nearly 9%," *New York Times*, April 26, 2010. As of February 20, 2012:
http://www.nytimes.com/2010/04/27/business/media/27audit.html

Poudenx, Pascal, "The Effect of Transportation Policies on Energy Consumption and Greenhouse Gas Emission from Urban Passenger Transportation," *Transportation Research Part A: Policy and Practice*, Vol. 42, No. 6, July 2008, pp. 901–909.

Pucher, John, and John L. Renne, "Rural Mobility and Mode Choice: Evidence from the 2001 National Household Travel Survey," *Transportation*, Vol. 32, No. 2, March 2005, pp. 165–186.

Reister, David B., and Warren D. Devine Jr., "Total Costs of Energy Services," *Energy*, Vol. 6, No. 4, April 1981, pp. 305–315.

Research and Innovative Technology Administration, "Mobility Services for All Americans," *Foundation Research Final Report*, Washington, D.C.: U.S. Department of Transportation, Intelligent Transportation Systems Joint Program Office, Federal Transit Administration, July 29, 2005. As of November 8, 2010:
http://www.its.dot.gov/msaa/msaa2/chapter3_4.htm

Ries, Charles P., Joseph Jenkins, and Oliver Wise, *Improving the Energy Performance of Buildings: Learning from the European Union and Australia*, Santa Monica, Calif.: RAND Corporation, TR-728-RER/BOMA, 2009. As of February 7, 2012:
http://www.rand.org/pubs/technical_reports/TR728.html

Roth, Matthew, "California's Personal Vehicle Sharing Law Could Diminish Need to Own a Car," *SF Streetsblog*, September 30, 2010. As of February 20, 2012:
http://sf.streetsblog.org/2010/09/30/californias-personal-vehicle-sharing-law-could-diminish-need-to-own-a-car/

Rydén, Christian, and Emma Morin, *Environmental Assessment Report Work Product 6,* Mobility Services for Urban Sustainability, January 18, 2005.

Santos, Adelia, Nancy McGuckin, Hikari Yukiko Nakamoto, Danielle Gray, and Susan Liss, *Summary of Travel Trends: 2009 National Household Travel Survey*, Washington, D.C.: Federal Highway Administration, FHWA-PL-11-022, June 2011. As of February 7, 2012:
http://nhts.ornl.gov/2009/pub/stt.pdf

Saunders, Caroline, Andrew Barber, and Greg Taylor, *Food Miles: Comparative Energy/Emissions Performance of New Zealand's Agriculture Industry*, Christchurch: Agribusiness and Economics Research Unit, Lincoln University, July 2006. As of October 6, 2011:
http://www.lincoln.ac.nz/Documents/2328_RR285_s13389.pdf

Schaller Consulting, *The New York City Taxicab Fact Book*, New York, March 2006. As of February 7, 2012:
http://www.schallerconsult.com/taxi/taxifb.pdf

Schuster, Thomas D., John Byrne, James Corbett, and Yda Schreuder, "Assessing the Potential Extent of Carsharing: A New Method and Its Implications," *Transportation Research Record,* Vol. 1927, 2005, pp. 174–181.

Shaheen, Susan A., Honda Distinguished Scholar in Transportation at the University of California at Davis and codirector of the Transportation Sustainability Research Center at the University of California at Berkeley, personal communication with the authors, November 28, 2011.

Shaheen, Susan A., Adam P. Cohen, and Melissa S. Chung, "North American Carsharing: 10-Year Retrospective," *Transportation Research Record*, Vol. 2110, May 2009, pp. 25–44.

Shaheen, Susan A., Adam P. Cohen, and Elliot Martin, "Carsharing Parking Policy: Review of North American Practices and San Francisco, California, Bay Area Case Study," *Transportation Research Record,* Vol. 2187, 2010, pp. 146–156.

Shaheen, Susan A., Adam P. Cohen, and J. Darius Roberts, "Carsharing in North America: Market Growth, Current Developments, and Future Potential," *Transportation Research Record,* Vol. 1986, 2006, pp. 106–115.

Silberglitt, Richard, Philip S. Antón, David R. Howell, Anny Wong, Natalie Gassman, Brian A. Jackson, Eric Landree, Shari Lawrence Pfleeger, Elaine M. Newton, and Felicia Wu, *The Global Technology Revolution 2020, In-Depth Analyses: Bio/Nano/Materials/Information Trends, Drivers, Barriers, and Social Implications*, Santa Monica, Calif.: RAND Corporation, TR-303-NIC, 2006. As of February 7, 2012:
http://www.rand.org/pubs/technical_reports/TR303.html

Sovacool, Benjamin K., "Conceptualizing Urban Household Energy Use: Climbing the 'Energy Services Ladder,'" *Energy Policy*, Vol. 39, No. 3, March 2011, pp. 1659–1668.

Steininger, Karl, Caroline Vogl, and Ralph Zettl, "Car-Sharing Organizations: The Size of the Market Segment and Revealed Change in Mobility Behavior," *Transport Policy,* Vol. 3, No. 4, October 1996, pp. 177–185.

Sustainable Energy Authority of Ireland, *Retro-Commissioning Review and User Needs Auditing*, Dublin, undated. As of February 7, 2012:
http://www.seai.ie/Your_Business/Large_Energy_Users/Special_Initiatives/Special_Working_Groups/Commercial_Buildings_Special_Working_Group_Spin_I/Retro-commissioning_and_User_Needs_Audit.pdf

Taylor, Cody, and Jonathan Koomey, "Estimating Energy Use and Greenhouse Gas Emissions of Internet Advertising," IMC², working paper, February 14, 2008.

Toman, Michael, James Griffin, and Robert J. Lempert, *Impacts on U.S. Energy Expenditures and Greenhouse-Gas Emissions of Increasing Renewable-Energy Use*, Santa Monica, Calif.: RAND Corporation, TR-384-1-EFC, 2008. As of February 7, 2012:
http://www.rand.org/pubs/technical_reports/TR384-1.html

U.S. Census Bureau, "Population Estimates," 2009a.

———, Population Division, "Annual Estimates of the Resident Population by Sex and Five-Year Age Groups for the United States: April 1, 2000 to July 1, 2009," *Population Estimates*, NC-EST2009-01, December 2009b.

———, Population Division, "Annual Estimates of the Resident Population for the United States, Regions, States, and Puerto Rico: April 1, 2000 to July 1, 2009," *Population Estimates*, NST-EST2009-01, December 2009c. As of June 11, 2010:
http://www.census.gov/popest/data/historical/2000s/vintage_2009/state.html

———, "Media Usage and Consumer Spending: 2003 to 2009," *Statistical Abstract of the United States*, 2011, Table 1130, p. 711. As of February 20, 2012:
http://www.census.gov/prod/2011pubs/11statab/infocomm.pdf

U.S. Energy Information Administration, *Annual Energy Review 2008*, Washington, D.C., DOE/EIA-0384(2008), June 26, 2009. As of February 7, 2012:
http://www.eia.gov/FTPROOT/multifuel/038408.pdf

———, "Energy Units and Calculators Explained," last updated November 22, 2011, referenced April 22, 2010. As of February 7, 2012:
http://tonto.eia.doe.gov/energyexplained/index.cfm?page=about_energy_units

———, *Annual Energy Outlook 2011*, Washington, D.C., DOE/EIA-0383(2011), April 26, 2011. As of February 7, 2012:
http://www.eia.gov/forecasts/archive/aeo11/index.cfm

———, *Annual Energy Outlook 2012 Early Release*, Washington, D.C., DOE/EIA-0383ER(2012), 2012. As of March 19, 2012:
http://www.eia.gov/forecasts/aeo/er/

U.S. Environmental Protection Agency, "Emission Facts: Average Carbon Dioxide Emissions Resulting from Gasoline and Diesel Fuel," Washington, D.C., EPA420-F-05-001, February 2005.

———, "eGRIDWeb," last updated February 7, 2012, referenced February 24, 2010. As of February 7, 2012:
http://cfpub.epa.gov/egridweb/

Vägverket, *Make Space for Car-Sharing!* Stockholm, Publication 2003: 88E, 2003.

Vringer, Kees, and Kornelis Blok, "The Energy Requirement of Cut Flowers and Consumer Options to Reduce It," *Resources, Conservation and Recycling*, Vol. 28, No. 1–2, January 2000, pp. 3–28.

Wear, Ben, "Smart Cars with a Case of the Stupids," *Austin American-Statesman*, June 14, 2010. As of November 8, 2010:
http://www.statesman.com/news/local/smart-cars-with-a-case-of-the-stupids-745487.html

Weber, Christoph, and Adriaan Perrels, "Modelling Lifestyle Effects on Energy Demand and Related Emissions," *Energy Policy*, Vol. 28, No. 8, July 2000, pp. 549–566.

Witkin, Jim, "Peer-to-Peer Car Sharing Gains Traction in Oregon," *New York Times*, June 9, 2011a. As of November 30, 2011:
http://wheels.blogs.nytimes.com/2011/06/09/peer-to-peer-car-sharing-gains-traction-in-oregon/

———, "Pod Cars, Moving Silently at Heathrow's Terminal 5," *New York Times*, August 5, 2011b. As of February 7, 2012:
http://wheels.blogs.nytimes.com/2011/08/05/pod-cars-moving-silently-at-heathrows-terminal-5/

Zhao, David, "Carsharing: A Sustainable and Innovative Personal Transport Solution with Great Potential and Huge Opportunities," Frost and Sullivan, January 28, 2010. As of February 20, 2012:
http://www.frost.com/prod/servlet/market-insight-top.pag?docid=190795176

Zipcar, "Commuters Across the U.S. and Canada Commit to Zipcar's 'Low-Car Diet' Challenge," press release, July 27, 2011. As of February 7, 2012:
http://zipcar.mediaroom.com/index.php?s=43&item=235